COVER PHOTOS

These NASA U-2 false color infrared photographs of Mt. St. Helens were taken from an altitude of 65,000 feet. Healthy vegetation appears red, ash is light blue, and snow and clouds are white. The top photo, taken 17 days before the May 18, 1980 eruption, shows tracts of mature timber and a small central crater. In the bottom photo, taken about a month after the event, ash and mudflows have extensively altered the surrounding terrain. See p. 285 for details of this dramatic volcanic event.

EXPLORING GEOLOGY

Introductory Laboratory Activities

Shannon O'Dunn, M.S.
Grossmont College, El Cajon, California

and

William D. Sill, Ph.D.
University of Texas, Austin

PRENTICE HALL, ENGLEWOOD CLIFFS, NEW JERSEY 07632

Copyright 1980, 1986
PRENTICE-HALL, INC.
A Division of Simon & Schuster
Englewood Cliffs, N.J. 07632

Formerly published as
General Geology of the Western United States

All rights reserved. No part of this publication may be reproduced or transmitted in any form or by any means, electronic or mechanical, including photocopy, recording, or any information storage and retrieval system, without permission in writing from the publisher.

10 9 8 7 6 5 4 3

ISBN 0-13-295668-3

Manufactured in the United States of America

Acknowledgments

We wish to thank all those persons and organizations whose contributions have made publication of this new edition possible. Carol Lawson, John Newhouse, Jacquelyn Freeberg, Mike Moore and others at the United States Geological Survey in Menlo Park were most helpful in assembling maps, charts, and photographs. The Grand Canyon Natural History Association kindly permitted the use of their geologic map of the Canyon. Thanks also to A. M. Bassett for his contributions to prior editions.

John Holden's exceptional illustrations are a valuable contribution to this new edition. Our gratitude to NASA Ames Research Center for providing us the outstanding infrared photographs of Mt. St. Helens for use on the cover. Barbara Clark's cover design using the infrared photos is well-deserving of special note.

To Kurt Servos of Menlo College, our grateful appreciation for his editorial assistance and timely suggestions.

Finally, a special word of appreciation goes to all of those geology instructors who responded to our request for information on how to further improve the manual—making it more useful to them and more appropriate for their students.

The Authors

To the Student

It all starts with the Earth. The ground we walk on, the fuel for heating and air conditioning, the silicon chips of computers, the synthetic fibers of our clothing, the metals and plastics of vehicles and machines, even the food we eat—all come from the rocks. Someone has to find the right rocks to begin the production of the world's wealth. Geology is the science of rocks and Earth processes. By definition it is a broad spectrum science that includes mineralogy, paleobiology, geophysics, geochemistry, and even planetary geology (one of the lunar astronauts was a geologist), and a number of unique specialized fields such as gemology, hydrogeology, petroleum geology, and geochronology, to name just a few.

Your progress through this manual will retrace the evolution of the geological sciences. Three hundred years ago geology was largely a descriptive discipline which centered on the identification and genesis of rocks, minerals, and fossils. The industrial revolution brought new pressure to bear upon geology to provide raw materials, such as metals and fuels. This necessitated accurate and readable maps portraying land shapes, the distribution of different rock types and ages, and the geometry of deformed rock bodies. The increased sophistication of ideas and instrumentation in this century has greatly altered the method of such protrayals, but the fundamental necessity of displaying a three-dimensional world on a two-dimensional plane remains an essential problem of geological study for the beginning student and practicing geologist alike.

You may decide to choose a career in one of the specialized fields of geology; or to take up rock collecting as a hobby; or to apply what you have learned to other interests, such as hiking, rock climbing, camping, or travel in general. This introductory course will help you participate more effectively as an informed citizen in the sensible development of geological resources and land use, and to view your surrounding landscape with a new awareness of the geological underpinnings of the biosphere.

Whatever your reason for studying geology, be willing to pay the price. There are basics in geology that, like the alphabet, must be learned in order to read the rocks. The basics begin with minerals and rock types—the fundamental units from which the Earth is constructed.

Finally, this manual is meant to be used—write in it, tear out pages, cut up diagrams, use it in any way that will help you enjoy and understand the Earth and its processes.

The Authors
Summer 1986

Contents

Chapter 1: Minerals 1
Definition 1
Physical Properties 2
Mineral Classification 6
Descriptive Mineral Tables 6
Review Questions 21
More Challenging Questions 23
Mineral Identification Quiz #1 25
Mineral Identification Quiz #2 27

Chapter 2: Rocks and Their Origins 29
Definition 29
Classification 29
Igneous Rocks 30
Sedimentary Rocks 39
Metamorphic Rocks 45
Rock Families and the Dynamic Lithosphere 48
Review Questions 51
More Challenging Questions 53
Rock Identification Quiz 55

Chapter 3: Aerial Photographs 57
Making Aerial Photographs 57
Using Aerial Photographs 58
Aerial Photographs and Topographic Maps 59
Advances in Aerial Photography 59
Selected Stereophoto Pairs 60
Review Questions 67
Aerial Photo Activities 69

Chapter 4: Fossils and Geologic Time 71
Fossils and Their Classification 71
Basic Principles of Stratigraphy 73
The Geological Time Scale 73
Exercises in Historical Geology Interpretation 73
Exer. #1: Age Estimates 83
Exer. #2: Correlation of Rock Units 85
Exer. #3: Composite Stratigraphic Columns 87
Exer. #4: Constructing a Geologic Map Explanation 89
Review Questions 91
Fossil Identification Quiz #1 93
Fossil Identification Quiz #2 95

Chapter 5: Topographic Maps 97
Map Types 97
Map Projections 97
Map Grids 98
Map Scales 102
Contour Lines 102
U.S. Geological Survey Topographic Quadrangles 103
Data from Topographic Maps 104
Exer. #1-#2: Cayucos, CA Quadrangle 109
Exer. #3-#4: Refuge, ARK-MISS Quadrangle 110
Exer. #5: Ennis, MT Quadrange 112

Exer. #6: Cordova, AK Quadrangle 114
Exer. #7: Lavic, CA Quadrangle 116
Exer. #8: Constructing a Topographic Map and Profile 133
Review Questions 137
More Challenging Questions 139
Topographic Map Symbols 141

Chapter 6: Structural Geology 143
Forces Producing Structural Deformation 145
Strike and Dip 146
Types of Structural Deformation 147
Exer. #1: Strike and Dip 157
Exer. #2: Folds 159
Exer. #3: Faults 163
Exer. #4: Fossils and Structural Interpretation 165
Exer. #5: Geologic History 167
Exer. #6: Structure and Geologic History 169
Exer. #7: Dip-Slip Faulting and Unconformities 171
Exer. #8: Thrust Faulting 173
Exer. #9: Geologic History of the Grand Canyon 175
Review Questions 177
More Challenging Questions 179

Chapter 7: Geologic Maps 181
Rock Unit Symbols and Contacts 181
Structure Symbols 184
Peripheral Information on Geologic Maps 184
Outcrop Patterns on Geologic Maps 184
Constructing a Geologic Cross-Section 188
Exer. #1: Constructing a Cross-Section from a Geologic Map 189
Exer. #2: Reading a Geologic Map 191
Exer. #3-#5: Grand Canyon 195
Exer. #6: Mapping Contacts from a Photograph 205
Exer. #7: Llano Uplift 207
Review Questions 213

Chapter 8: Plate Tectonics 215
Plate Boundaries 216
Mantle Plumes and Triple Junctions 217
Exer. #1: World Ocean Physiography 219
Exer. #2: The North Pacific 222
Exer. #3: Kilauea Crater 225
Exer. #4: The Cascade Range 228
Exer. #5: Afar Triangle 230
Exer. #6: San Andreas Fault 234
Review Questions 249

Chapter 9: Applied Geology 251
Exer. #1: Economic Deposits in the Stable Central Rockies:
  Devil's Tooth, WY 253
Exer. #2: Geological Resources: Origin and Occurrence of Petroleum 259
Exer. #3: Ground Water 263
Exer. #4: Development of a Marine Recreational Harbor: Mission Bay
  Park, CA 267
Exer. #5: Urban Landsliding: The Portuguese Bend Landslide 273
Exer. #6: Tectonic Hazards: Locating Earthquake Epicenters 279
Exer. #7: Mount St. Helens, WA 285

Index 291

Minerals 1

INTRODUCTION

Most people are familiar with the commonly applied classification of all natural entities into the three categories of animal, vegetable, and mineral. We now know that these subdivisions are not in the strictest sense discrete. Some organisms exist which combine animal locomotion with plant photosynthesis. Others, such as viruses, grow and reproduce under some conditions, and crystallize like minerals under other conditions. However, the general implications of this fundamental breakdown are valid; minerals are *non-living*, and occur in many *species* as do animals and plants.

The study of minerals is the area of geology most commonly translated into a satisfying hobby for the layman. Indeed, one need not be a full-fledged geologist to appreciate the aesthetic and intrinsic value of minerals, and to enjoy identifying and collecting them. The primary objective of this chapter is to introduce the most common minerals selected from a kingdom of several thousand species, and to define the various physical criteria used to distinguish one mineral species from another.

DEFINITION

In order to be considered a mineral, a substance must satisfy four criteria: *it must be naturally occurring, inorganic, have a definite chemical make-up, and be crystalline in nature.* The implications of these criteria are discussed below:

Naturally Occurring: Rules out man-made substances. While it is possible to "grow" materials in the laboratory which might be indistinguishable from a natural specimen, these are not considered minerals. The only important application of this rule would be in relation to synthetically produced rubies, sapphires, emeralds, or other gems. These may be superior in appearance to nature's often-flawed examples of the same material, but the synthetic gems have no investment value.

Inorganic: This means that minerals are not, and never were, living. Minerals which contain carbon are the only "gray area" here. Diamonds (C) *may* have an organic background; graphite (C) almost certainly does; and calcite ($CaCO_3$), the material of clam shells and coral reefs, is definitely a *product* of organic processes. However, these are accepted as minerals by convention.

Definite Chemical Make-up: The basic identity of a mineral is in its chemistry. Each species contains specific elements in definite proportions. SiO_2 is the chemical formula for the most common continental crust mineral species, quartz. This formula states that any crystal or fragment of quartz has one silicon atom for every two oxygen atoms. Both FeS and FeS_2 are valid minerals, but they are not the *same* mineral. A few minerals do have defined and acceptable limited substitution of one element for another; this will be indicated by ( ) in the chemical formula.

Crystalline: In crystalline materials, the component atoms are arranged in an orderly network at specific distances and angles to one another. This property may be demonstrated by x-ray analysis. Megascopically, crystallinity is shown in the external symmetry of crystals, and the consistent breakage patterns of many species. Quartz and volcanic glass both satisfy the first three mineral criteria; but only quartz is crystalline, and therefore a mineral. No *glass* is a mineral because these types of materials, like liquids, lack a definite atomic arrangement.

PHYSICAL PROPERTIES

Each mineral species has a discrete chemistry, and therefore can be identified by chemical tests. Since this approach is not very practical for field work, mineralogists have established a series of physical properties which will suffice to distinguish the more common minerals from one another. These properties are listed and described below in order of frequency of application.

In mineral identification not all of these properties apply to, or are important for, each species. Some minerals have one specific property which serves to identify them quickly (hardness of diamond, grooves on plagioclase), while others can be recognized by a combination of features (fracture and hardness of quartz, luster and color of sulfur). Students should understand all the physical properties as defined, and recognized the minimum number of specific properties needed for identification of each mineral to be learned.

    I. **HARDNESS:** This is a measure of the resistance of a mineral to scratching, and is evaluated on a relative scale of 1 to 10. Lower numbers indicate less resistance to scratching.

> **Mohs' Hardness Scale**
> 1: Talc
> 2: Gypsum
>   (2¼: fingernail)
> 3: Calcite
>   (3½: copper coin)
> 4: Fluorite
> 5: Apatite
>   (5-5½: steel nail, knife)
>   (5½-6: glass)
> 6: Orthoclase
> 7: Quartz
> 8: Topaz
> 9: Corundum
> 10: Diamond

Each mineral on this scale can scratch a mineral with an equal or lower number, but none with a higher number. For example, calcite can scratch calcite, your fingernail, gypsum, or talc, but nothing with a hardness greater than 3. All minerals have a hardness either equal to or between the minerals listed on this scale. Pyrite ("fool's gold"), for example, has a hardness of 6, the same as orthoclase; sphalerite can scratch calcite but not fluorite, and so has a hardness of 3½. Minerals with a hardness of less than 6 will not withstand prolonged wear as gemstones.

Use considerable pressure in testing for hardness. Make a small scratch about ¼-inch long to minimize defacing the samples. Wipe any apparent scratches with your finger to be sure the mark is genuine, and not just a powder streak from the softer mineral. When using the glass test plate, place it flat on the laboratory table to avoid injury.

II. **LUSTER:** The appearance of a mineral's surface in reflected light.

Metallic Luster: These minerals have a "metallic glint" and reflect light effectively; they are opaque and relatively heavy.

Non-metallic Luster: The most common are:

Adamantine = brilliant, like that of a polished diamond.
Vitreous = glassy, like glazed porcelain or quartz.
Resinous = like resin; sphalerite is an example.
Pearly = similar to that created by transparent microscopic scales in natural pearls; talc, some gypsum.
Silky = resulting from parallel fibers; asbestos, "satin spar" gypsum
Earthy = dull, little reflection; kaolinite, goethite.

III. **COLOR:** This would seem to be a simple attribute to deal with, and yet it really does not work very well for mineral identification. Colors are subjective to the observer, and unreliable for many specimens. Quartz, fluorite, calcite, and other fundamentally colorless minerals may take on the full spectrum of hues depending on trace element contamination. Also, small crystals usually look lighter in aggregate. In some specimens a thin surface coating of exotic material may be misleading. Color is most dependable for the metallic minerals.

IV. **STREAK:** This is the color of a mineral when powdered. Streak is generally constant even when the color of uncrushed material varies. Most non-metallic minerals have a non-diagnostic white streak, but metallic samples have a dependable streak. Specimens may be crushed in a mortar, but the usual practice is to drag them across the abrasive surface of an unglazed tile. The latter approach is useless with substances which are harder than the streak plate (5½ to 6).

V. **CLEAVAGE:** This is the tendency of a mineral to split along certain planes determined by internal atomic arrangement and bonding strength. Cleavage planes, produced by rupture in some minerals, should be distinguished from crystal faces. Crystal faces are the surfaces bounding a crystal; the natural shape a mineral takes on when room for growth is present. Calcite, for example, may *grow* in elongated six-sided pyramids or a variety of other shapes, but always *cleaves* into a rhombohedron. Rhombohedrons have faces shaped like parallelograms (△). Quartz has no cleavage. Halite (table salt) and galena characteristically break into cubes, or portions thereof, and thus have cubic cleavage.

VI. **FRACTURE:** This is the non-planar breakage of minerals; i.e., no well-defined cleavage. Fracture may be described as uneven (rough), splintery, or conchoidal (curved breaks typical of glass). Quartz has conchoidal fracture.

VII. **CRYSTAL FORM:** Most minerals will take on a definite external geometric form when unobstructed growth is allowed. This form is an expression of the orderly internal atomic arrangement, or crystallinity. Whole crystals and crystal faces are therefore an aid to identification. A complete study of crystallography is beyond our scope here. Table 1-1 presents the basic crystal systems and gives common examples of each.

VIII. **MISCELLANEOUS:** Other properties which are less common or even unique to certain minerals include:

Natural magnetism (magnetite)
Feel—talc feels slippery; halite feels greasy
Specific gravity—galena is dense; halite is not
Effervescence in acid—calcite reacts with cold weak hydrochloric acid (HCl)
Tenacity—reaction under stress; brittle (galena), flexible (talc), elastic (muscovite), malleable (native copper)
Structure—common mode of occurence; crystalline (garnet), granular (olivine), fibrous (asbestos)
Double refraction—transparent calcite transmits a split image

Fluorescence—emission of color under ultraviolet light (fluorite, some calcite, some diamonds, beryl)
Odor—moist kaolinite gives off earthy odor
Crystal arrangement—closely spaced, fine parallel grooves ("record grooves") on plagioclase feldspar

Table 1-1. The Six Crystal Systems

All mineral species fall into one of the six crystal systems shown below. Careful study of the examples will help you to use crystal form as an identification tool for complete crystals and fragmented or partially developed specimens.

Crystal System	Common Forms and Examples		
Cubic	Cube Ex: Halite	Octahedron Ex: Fluorite, native gold	Dodecahedron Ex: Garnet
Hexagonal	Hexagonal prism with 1 termination Ex: Quartz	Hexagonal plates Ex: Molybdenite	Rhombohedron Ex: Calcite
Tetragonal	Tetragonal disphenoid Ex: Chalcopyrite		Tetragonal prism with 2 terminations Ex: Scheelite
Orthorhombic	Combined form Ex: Topaz	Combined form Ex: Staurolite twin	Orthorhombic prism Ex: Sulfur
Monoclinic	Combined form Ex: Gypsum	Combined form Ex: Hornblende	Combined form Ex: Orthoclase
Triclinic	Combined form Ex: Rhodonite	Twinned plagioclase Ex: Albite	Combined form Ex: Plagioclase feldspar

6 Exploring Geology

MINERAL CLASSIFICATION

Minerals may be categorized as *ore-forming* or *rock-forming*. This nonexclusive breakdown is practical rather than theoretical.

Ore minerals may be metallic or non-metallic, and historically they have been mined for profit. There are numerous species of ore minerals, but taken together they make up only a small percentage of the Earth's crust. An ore is any natural substance that can be extracted from the earth and sold commercially.

There are relatively few rock-forming minerals, but they comprise more than 95 percent of the Earth's crust, both continental and oceanic, plus the bulk of the upper mantle below the crust. These minerals fall into eight families, each with several species. A mineral family is one in which the general structural arrangement of the atoms is fixed, but substitution of one similarly-sized atom for another varies the chemical composition and produces minor differences in overall physical properties.

The majority of the rock-forming minerals are **silicates**, minerals which are based on a primary network of silicon and oxygen with or without the addition of one or more metallic elements. The principal rock-forming minerals are:

Description	Mineral Species	Family
Ferromagnesian minerals. High in iron and magnesium; heavy, dark. Common in **mafic** igneous rocks.	Olivine	Olivine series
	Augite	Pyroxene
	Hornblende	Amphibole
	Biotite	Mica
Non-ferromagnesian minerals. High in light metals as aluminum, potassium, sodium, calcium. Light in color and specific gravity. Common in **felsic** igneous rocks.	Muscovite	Mica
	Plagioclase	Feldspar
	Orthoclase	Feldspar
	Quartz	Quartz
Species range from ferromagnesian to non-ferromagnesian. Found in felsic and mafic igneous rocks.	Andradite, etc.	Garnet
Minerals formed at the Earth's surface (i.e., products of weathering, evaporation, etc.).	Kaolinite	Clay
	Calcite	Salt
	Gypsum	Salt
	Halite	Salt

DESCRIPTIVE MINERAL TABLES

The Descriptive Mineral Tables (pp. 7-13) detail the most diagnostic features of more than fifty ore- and rock-forming minerals. The tables separate mineral species into two luster categories, Metallic (p. 7) and Non-Metallic (p. 8). Within each of these categories, species are arranged in order of increasing hardness. Streak, which is an especially useful diagnostic property of the minerals with metallic luster, is noted for these species, whereas specific luster types are given for the non-metallic minerals. Unique or notable characteristics of each mineral which may aid in identification are detailed under Conspicuous Features.

Flow Sheets for keying out minerals described in this chapter will be found on pages 17-19. The chart on page 17 will aid in identifying minerals with metallic luster. The charts on pages 18 and 19 cover minerals with non-metallic luster. For most effective use of the flow sheets, students should determine streak for metallic minerals, and a hardness estimate for non-metallic ones.

DESCRIPTIVE MINERAL TABLE
I. Minerals with Metallic Luster

Name and Composition	Hardness	Color	Streak	Conspicuous Features	Crystal System
Graphite C Elemental carbon	1	Silver-gray	Black	Marks paper like a pencil, greasy feel, light in weight. One perfect cleavage.	hexagonal
Molybdenite MoS_2 Molybdenum sulfide	1-1½	Bluish-gray	Greenish-gray	Soft, flexible, shiny plates (one perfect cleavage), often with hexagonal outline. Marks paper.	hexagonal
Galena PbS Lead sulfide	2½	Silver-gray	Black	Cubic or octahedral crystals, bright metallic luster, heavy. Cubic cleavage.	cubic
Native Copper Cu	2½-3	Copper-rose	Copper-rose	Copper-rose color on fresh surfaces. Heavy and malleable. Rare crystals; usually in compact masses. Often has a pale green surface coating of malachite.	cubic
Native Gold Au	2½-3	Gold, white-gold, rose-gold	Same as color	Color varies with impurities. Extremely heavy. May be extended or shaped by hammering or rolling (malleable). Dissolves only in aqua regia. Rare small crystals, and dendrites (tree-like growths); nuggets in sedimentary deposits.	cubic
Native Silver Ag	2½-3	Silver-white	Silver-white	Usually tarnished dark gray. Irregular fracture. Very heavy. May be gouged or sliced with a knife (sectile). May occur as dendrites (see Gold) and wires in calcite and other minerals.	cubic
Bornite Cu_5FeS_4 Copper iron sulfide	3	Rose to brown	Gray-black	"Peacock ore." Iridescent alteration coating common; brittle; conchoidal fracture.	cubic
Chalcopyrite $CuFeS_2$ Copper iron sulfide	3½	Dark brass-yellow	Greenish-black	Often tarnished iridescent greenish blue. Brittle, fairly soft, usually massive. Conchoidal fracture.	tetragonal
Pyrite FeS_2 Iron disulfide	6	Light brass-yellow	Black	Occurs in cubes with grooved faces, and pyritohedrons with 5-sided faces. Called "fool's gold," much lighter than true gold. Poor cleavage; fragile.	cubic

8 Exploring Geology

Name and Composition	Hardness	Color	Streak	Conspicuous Features	Crystal System
Magnetite Fe_3O_4 Iron oxide	6	Black	Black	Magnetic, granular or octahedral crystals common. No cleavage.	■
Specular Hematite Fe_2O_3 Iron oxide	6	Shiny steel-gray	Dark red	Glittering flakes or wavy sheets. Streak is distinctive. Tendency to flake obscures true hardness.	⬢

II. Minerals with Non-Metallic Luster

Name and Composition	Hardness	Color	Luster	Conspicuous Features	
Talc $Mg_3Si_4O_{10}(OH)_2$ Hydrous magnesium silicate	1	White, pale green	Pearly	Extremely soft; soapy feel. Impurities may increase apparent hardness. Often in scaly masses. One perfect cleavage.	⬢
Native Sulfur S	1½-2½	Yellow	Resinous, greasy	Yellow color, low hardness, light weight. Detectable sulfur odor. Often in well-developed blocky crystals, or as a fine coating on volcanic rock.	⬢
*Gypsum $CaSO_4 \cdot 2H_2O$ Hydrous calcium sulfate	2	Colorless, white; sometimes pale orange	Vitreous, pearly	Selenite is clear, satin spar is fibrous, alabaster is massive. Selenite may occur in large (to 3 m) sword-like crystals; or in bladed groups incorporating sand and known as "desert roses." One perfect cleavage.	⬢
Borax $Na_2B_4O_7 \cdot 10H_2O$ Hydrous sodium borate	2	White	Vitreous	Short, stubby crystals. Brittle, soft. Also in earthy, massive form. Conchoidal fracture.	⬢
Chlorite Hydrous aluminous ferromagnesian silicate	2	Light to dark green	Vitreous to earthy	Green color and micaceous habit (one good cleavage). Flakes are not elastic like mica.	⬢
*Kaolinite $Al_2Si_2O_5(OH)_4$ Hydrous aluminosilicate	2-2½	White, cream	Earthy, dull	Soft, powdery texture. Smells earthy when damp. Usually in clay-like masses with dull appearance.	⬢

Name and Composition	Hardness	Color	Luster	Conspicuous Features	Crystal System
Cinnabar HgS Mercury sulfide	2-2½	Cinnamon red	Adamantine to dull	Color diagnostic. May appear almost metallic or in earthy, pinkish-red masses. Scarlet streak. Toxic.	
Carnotite $K_2(UO_2)_2(VO_4)_2 \cdot 3H_2O$ Hydrous potassium uranium vanadate	Very soft	Canary yellow	Dull, earthy	Usually a coating or powder in sandstone or other rock; imparts a strong yellow color. Very radioactive. Hardness indeterminate.	
***Biotite Mica** $K(Mg,Fe)_3 AlSi_3O_{10}(OH)_2$ Hydrous potassium aluminum ferromagnesian silicate	2½	Dark brown, black	Vitreous	Black mica. Occurs in six-sided mica "books" and as scattered flakes. Peels into thin elastic greenish-brown sheets along one perfect cleavage.	
***Muscovite Mica** $KAl_2(AlSi_3O_{10})(OH)_2$ Hydrous potassium aluminum silicate	2½	Colorless, pale green	Vitreous to pearly	White mica. Occurs in mica "books" and as scattered flakes. Peels into thin elastic transparent sheets along one perfect cleavage.	
Lepidolite Mica $KLi_2(AlSi_4O_{10})(OH)_2$ Hydrous potassium lithium aluminum silicate	2½-4	Colorless, lilac, yellow	Vitreous to pearly	Lilac color is diagnostic. Often in granular masses of small mica "books." Lavender mica.	
***Halite** $NaCl$ Sodium chloride	2½	Colorless, salmon, pastels	Vitreous to greasy	Easily dissolves in water. Often has stepped-down "hopper" faces. Crystal masses or coatings on other material. Cubic cleavage.	
Asbestos $Mg_6Si_4O_{10} \cdot (OH)_8$ Hydrous magnesium silicate	2½-3	Light green, light brown	Silky	Long, thread-like fibers with silky sheen. The commercial variety is fibrous serpentine.	

10 Exploring Geology

Name of Composition	Hardness	Color	Luster	Conspicuous Features	Crystal System
**Calcite* $CaCO_3$ Calcium carbonate	3	Colorless, white; rarely pastels	Vitreous	Effervesces freely in cold dilute hydrochloric acid. Doubly refracting. Frequently in rhombohedral crystals; hundreds of other forms known. May be fluorescent. Perfect rhombohedral cleavage.	hexagonal
Barite $BaSO_4$ Barium sulfate	3	Colorless, white, blue	Vitreous	Heavy for a non-metal. Often occurs as tabular crystals; such crystals in circular arrangement form "barite roses." Two cleavages.	orthorhombic
Bauxite Hydrous aluminum oxide mixture	3-3½	White; usually stained with iron oxide	Earthy	Pea-sized round concretionary grains show color banding in cream, yellow, and brown. Actually a rock made up of various hydrous aluminum oxides of which it is the only ore.	—
Sphalerite ZnS Zinc sulfide	3½	Usually yellow-brown; also black, green, red	Adamantine to metallic	Light yellow streak in most color varieties. Heavy. Occurs as crystals, compact masses, and coatings. Perfect dodecahedral cleavage; cleavage chunks often triangular in shape.	cubic
Azurite/ $Cu_3(CO_3)_2(OH)_2$ **Malachite** $Cu_2CO_3(OH)_2$ Hydrous copper carbonates	3½-4	Azure blue and bright green, respectively	Dull or velvety	Colors and association distinctive; both effervesce in hydrochloric acid. Azurite often in radiating masses. Malachite frequently in curved masses exhibiting color banding in shades of green.	monoclinic
Dolomite $CaMg(CO_3)_2$ Calcium magnesium carbonate	3½-4	White, yellow, pink	Vitreous to pearly	Slow effervesces in cold dilute acid when powered. Pale pink color is indicative. Often associated with calcite. Usually in rhombohedral crystals; perfect rhombohedral cleavage.	hexagonal
Fluorite CaF_2 Calcium fluoride	4	Colorless, all pastels, deep purple	Vitreous	Crystals often cubic or octahedral. Color banding common. Usually fluorescent in ultraviolet light. Octahedral cleavage.	cubic
Colemanite $Ca_2B_6O \cdot 5H_2O$ Hydrous calcium borate	4½	Colorless, white	Vitreous	May be in stubby, glassy crystals, or in compact granular masses. One perfect cleavage.	monoclinic
Apatite $Ca_5(PO_4)_3F$ Calcium fluoro-phosphate	5	White, blue, brown; also green, yellow	Vitreous	Will not scratch glass. Commonly in 6-sided prisms. One poor cleavage.	hexagonal

Name and Composition	Hardness	Color	Luster	Conspicuous Features	Crystal System
Scheelite $CaWO_4$ Calcium tungstate	5	White, yellow, brown	Vitreous	Will not scratch glass. Heavy. Fluoresces. Crystal faces may be grooved. One good cleavage.	
Earthy Hematite Fe_2O_3 Iron oxide	5	Dull brownish red to bright red	Sub-metallic to earthy	Characteristic red-brown streak. Often earthy and too powdery for accurate hardness test. May be granular or oolitic. Crystals rare; no cleavage.	
Goethite $HFeO_2$ Hydrous iron oxide	5-5½	Dark rusty brown, ochre yellow	Dull, earthy	Streak distinctive yellow-brown. Often spongy, porous or earthy; also bladed, fibrous. May occur in cubes and pyritohedrons as an alteration of pyrite. Crystalline form of limonite.	
Rhodonite $MnSiO_3$ Manganese silicate	6	Pink to deep rose	Vitreous	Massive, dense or granular aggregates often have black veins. Color and hardness diagnostic. Blocky crystals, nearly 90° cleavage.	
***Hornblende** Complex hydrous ferromagnesian silicate with calcium, aluminum, and sodium	5½-6	Greenish-black to black	Vitreous	Barely scratchs glass. Shiny on cleavage faces; opaque; often splintery at edges. Usually massive; occasionally in chunky crystals. Two cleavage angles: 124° and 56°.	
***Augite** Similar to hornblende, but anhydrous	6	Dark green to greenish black	Vitreous to dull	Opaque prismatic crystals. Usually duller and greener than closely related hornblende. Two cleavages (87° and 93°), and uneven fracture.	
***Orthoclase Feldspar** $KAlSi_3O_8$ Potassium aluminosilicate	6	White pink, salmon	Vitreous	Will scratch glass. Wavy internal pattern and pink color distinguish it from plagioclase when present. May be massive, or in large, well-developed coffin-shaped crystals. Two good cleavages.	
***Plagioclase Feldspar** $NaAlSi_3O_8 \leftrightarrow CaAl_2Si_2O_8$ Sodium and/or calcium aluminosilicate	6	White, gray	Vitreous	Two good cleavages. Will scratch glass. "Record grooves." Rectangular cleavage faces often seen in igneous rocks.	

12 Exploring Geology

Name and Composition	Hardness	Color	Luster	Conspicuous Features	Crystal System
Spodumene $LiAlSi_2O_6$ Lithium aluminum silicate	6½	Colorless, white, lavender	Vitreous	Elongated prismatic crystals. Associated with lepidolite, tourmaline, beryl. Deep grooves often parallel long crystal faces. Perfect prismatic cleavage.	▰
*Olivine $(Mg,Fe)_2SiO_4$ Magnesium iron silicate	6½-7	Olive green	Vitreous	Crystals often appear as glassy green beads, isolated or in masses. Color distinctive. Conchoidal fracture.	▰
Epidote $Ca_2(Al,Fe)_3Si_3O_{12}(OH)$ Hydrous calcium iron aluminum silicate	6½-7	Light to dark green	Vitreous	Usually a dull avocado green when massive; crystals are shiny dark green, with striations and well-developed cleavage.	▰
*Quartz Family SiO_2 Rock Crystal Milky Quartz Smoky Quartz	7	Colorless White Gray, brown	Vitreous to greasy	Crystals are 6-sided prisms, often with terminations and steps perpendicular to crystal length. Crystals may be in clusters, or line cavities in rock; some weigh several hundred pounds. Conchoidal fracture; no cleavage.	⬢
Rose Quartz Amethyst Citrine	7	Pink Purple Yellow			
Chalcedony (petrified wood, flint, chert, agate, jasper)	7	Variable colors	Waxy	Massive, dense, often bumpy masses; waxy surface. Color banded or mottled appearance common. Not wholly crystalline. May line rock cavities to form geodes, or replace organic material to "petrify" wood, shell or bone.	—
Staurolite $FeAl_4Si_2O_{10}(OH)_2$ Hydrous iron aluminum silicate	7-7½	Brown	Vitreous to dull	Usually found as prismatic crystals; often twinned to form crosses. Crystal faces are pitted and rough. Cruciform twinning is diagnostic when present.	▰
Tourmaline Complex borosilicate of aluminum, sodium, magnesium, iron, lithium	7-7½	Black, brown, green, pink, blue yellow	Vitreous	Typically in elongated crystals with grooved faces and rounded triangular cross section. Common variety shiny black. Crystals often occur in parallel or radiating groups. No cleavage.	⬢

Name and Composition	Hardness	Color	Luster	Conspicuous Features	Crystal System
Garnet $Fe_3Al_2(SiO_4)_3$ Iron aluminum silicate, also with calcium and magnesium	7-7½	Brown, red; also purple, green, yellow, black, pink	Vitreous to resinous	Commonly in shades of red. Dodecahedral crystals have diamond-shaped faces. Color and hardness aid identification. Transparent to opaque. No cleavage.	■
Beryl $Be_3Al_3Si_6O_{18}$ Beryllium aluminum silicate	7½-8	Colorless, white, pink, blue, light green, emerald green	Vitreous	Commonly pale green, and in 6-sided prisms with flat terminations. Harder than quartz. Pale blue variety is aquamarine; chromium green variety is emerald.	⬢
Topaz $Al_2SiO_4(OH,F)_2$ Hydrous fluoro-aluminum silicate	8	Colorless, white, golden yellow, light blue	Vitreous	Internal rainbows. Striations on crystal faces. Distinct glassy prismatic crystals with perfect basal cleavage exhibiting diamond-shaped cross section.	⬢
Corundum Al_2O_3 Aluminum oxide	9	Gray, all pastels, red, dark blue, brown	Vitreous to greasy	Commonly in barrel-shaped 6-sided crystals, tapered or with flat ends. Extremely hard. No cleavage.	⬢
Diamond C Elemental carbon	10	Colorless, pastels, brown, blue, yellow, gray, black, red, green	Adamantine to greasy	Octahedral crystals with greasy luster. Hardest known substance. Two directions of cleavage.	■

14 Exploring Geology

Muscovite from San Diego County, California. (R. H. Currier)

Plagioclase Feldspar. (Photo by E. Jonas.)

Fluorite cleavage pieces. (Photo by E. Jonas.)

Staurolite penetration twin from North Carolina. (R. H. Currier)

Gypsum, variety Satin Spar. (Photo E. Jonas.)

Photos 1a. Minerals

16 Exploring Geology

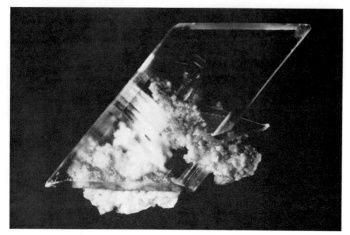

Gypsum, variety Selenite, Mexico. (R. H. Currier)

Orthoclase Feldspar twin, Norway. (R. H. Currier)

Blue tourmaline crystals. (U.S.G.S.)

Chrysotile asbestos in serpentine from Trinity County, California. (Photo by E. H. Pampeyan, U.S.G.S.)

Quartz crystals. (U.S.G.S.)

Photos 1b. Minerals

MINERALS WITH METALLIC LUSTER

MINERAL COLORS	STREAK	MINERAL	CHARACTERISTICS
Light silver gray	Silver	Native Silver	Light color, soft
	Black	Galena	Shiny cubes
Dark silver gray	Greenish-gray	Molybdenite	Flexible plates
	Dark red	Specular Hematite	Streak
	Black	Graphite	Marks paper
Black	Black	Magnetite	Magnetic
Golden	Golden	Native gold	Soft
	Greenish-black	Chalcopyrite	Iridescent tarnish
	Black	Pyrite	Brassy cubes, H = 6
Rose	Copper	Native copper	Color diagnostic
	Gray-black	Bornite	"Peacock" tarnish

MINERALS WITH NON-METALLIC LUSTER—Hardness ≤ 5½ (Will not scratch glass)

HARDNESS 1 → 10	MINERAL COLORS	MINERALS	CHARACTERISTICS
Can be scratched by fingernail	White, colorless	Talc	Soapy feel
		Kaolinite	Chalky appearance, earthy odor
		Gypsum	Satiny fibers or one good cleavage, light weight
		Borax	Conchoidal fracture
	Yellow	Native sulfur	Streak for rotten egg odor
		Carnotite	Disseminated powder, radioactive
	Green	Chlorite	Flakes (mica)
Not scratched by fingernail / Won't scratch calcite	White, colorless	Muscovite	One perfect cleavage (mica)
		Halite	Cubic forms
	Pale green	Asbestos	Fibrous character
	Lavender	Lepidolite	Scales or flakes (mica)
	Dark red	Cinnabar	Cinnamon color
	Black	Biotite	One perfect cleavage (mica)
Will scratch calcite / Won't scratch fluorite	White, colorless	Calcite	Rhombohedral forms, effervesces in HCl
		Barite	Heavy
		Dolomite	Effervesces with HCl on scratches
	Brown	Bauxite	Spherical structures
		Sphalerite	Yellow streak
	Bright blue, green	Azurite	Azure blue
		Malachite	Green, banded
Nail will scratch / Will scratch fluorite	White, colorless	Colemanite	Low density
		Scheelite	Heavy, may fluoresce
	Pastels	Fluorite	Octahedral cleavage, often color banded
	Green	Apatite	Often in stubby six-sided crystals
	Brown, reddish brown	Goethite	Yellow-brown streak
		Hematite	Red-brown streak

18 Exploring Geology

MINERALS WITH NON-METALLIC LUSTER—Hardness ≥ 5½ (Will scratch glass)

HARDNESS 1 → 10	MINERAL COLORS	MINERALS	CHARACTERISTICS
Won't scratch quartz	Colorless, pastels →	Spodumene	Splintery cleavage
	White to salmon →	Orthoclase	Blocky cleavage
	White to gray →	Plagioclase	Parallel "record" grooves
Nail won't scratch	Deep pink →	Rhodonite	Dark streaks
	Green {	Augite	Dark green to black
		Olivine	Glassy "beads"
		Epidote	Earthy pistachio color or shiny dark green crystals
	Black →	Hornblende	Vitreous luster; splintery cleavage, elongate crystals
Will scratch quartz	Colorless, white, variable →	Quartz family	See page 12 for description of varieties
	Colorless to transparent peach →	Topaz	One perfect cleavage, internal rainbows
	Colorless →	Diamond	Octahedral forms, "frosted" crystals
	Opaque pale green →	Beryl	Poor cleavage, scratches quartz
	Translucent pink or green; black →	Tourmaline	Elongate crystals with grooves
	Light brown → {	Staurolite	Elongate crystals
		Corundum	Stubby hexagonal prisms
	Dark brown or red →	Garnet family	No cleavage, equidimensional crystals, diamond shaped faces

Minerals 21

Chapter 1: Minerals   Name Benny Chong

Section _____

REVIEW QUESTIONS

1. Indicate whether or not the following materials are minerals. For those which do not qualify, briefly explain *why* they are not minerals.

   yes/no

   a. snowflake yes

   b. coal no, organic background (rotten plants, animals)

   c. table salt yes

   d. window glass no, lacks definite atom arrangement.

   e. graphite yes

2. If diamond is the hardest known substance, what mineral substance could be used to cut and polish it? Other diamonds and synthetic ones.

3. How would you distinguish pyrite ("fool's gold") from native gold?
   Pyrite would be heavier and harder (6). It is very brittle, not malleable like native gold and exhibits a definite black streak.

4. Citrine is often sold as topaz. What differences between these two minerals could be used to tell them apart? Topaz is harder (8) and has internal rainbows. It also has perfect basal cleavage whereas citrine doesn't.

5. Why isn't clear, sapphire-blue azurite used as a faceted gem in jewelry?
   Azurite has a hardness of $3\frac{1}{2}$ to 4, and anything below a 6 will not withstand prolonged wear as gemstones. It also reacts with HCl.

6. What are the two most important minerals of the feldspar family? Include chemical formulas.
   Plagioclase: $NaAlSi_3O_8$ or $CaAl_2Si_2O_8$
   Orthoclase: $KAlSi_3O_8$

7. What color streak would olivine give on an unglazed tile? White streak

   Why? Olivine is harder than the tile, unable to leave a powder.

8. What common physical property do the mica minerals share? Cleavage

9. How would you distinguish between orthoclase and plagioclase feldspar by sight identification?
   Orthoclase is pinkish and a wavy internal pattern. Plagioclase is white to grey and "record grooves" running down its rectangular cleavage faces.

10. What is the simplest silicate chemical formula? __SiO$_2$__

11. Name two minerals abundant in light-colored, lightweight igneous rocks:

    __Plagioclase__   __Orthoclase__

12. How can you physically distinguish between the closely related species augite and hornblende?

13. What do amethyst, petrified wood, and flint have in common? __Belong to Quartz Family (Hardness: 7)__

14. For what group of minerals is color a fairly useful identification tool? __Metallic__

Chapter 1: Minerals

Name _Benny Chong_

Section _____

MORE CHALLENGING QUESTIONS

1. What are the advantages and disadvantages of a synthetic gemstone to the buyer?

2. Why is hematite the most important ore of iron, although magnetite contains a higher percentage of iron?

3. Define mineral *cleavage*.

4. Give two examples of minerals which are precipitated by the evaporation of seawater:

   _____   _____

5. List three specific distinctive properties for each of these minerals:

Calcite	*Galena*	*Sulfur*

6. What properties should a mineral possess in order to be considered a desirable and valuable gemstone?

7. Name two minerals which characterize dark, heavy (simatic) igneous rocks:

   _____   _____

8. Name each of the minerals described below:
   a. the hardest _Diamond_
   b. the most common individual mineral *species* _Quartz_
   c. clay mineral _____
   d. gemstone emerald _____
   e. the most abundant mineral *family* _____
   f. magnetic _Magnetite_
   g. "peacock ore" _Bornite_
   h. non-metal native element (name two) _Sulfur_
      Graphite

i. an ore of uranium _____

j. aluminum ore _____

k. fibrous serpentine _____

l. mineral constituent of bone _____

9. What is the chemical difference between pyrite and chalcopyrite?

How can you physically tell them apart?

10. Name four minerals that are green gemstones:

_____   _____

_____   _____

Minerals 25

Name _____

Section _____

MINERAL IDENTIFICATION QUIZ #1

1. _____
2. _____
3. _____
4. _____
5. _____
6. _____
7. _____
8. _____
9. _____
10. _____
11. _____
12. _____
13. _____
14. _____
15. _____
16. _____
17. _____
18. _____
19. _____
20. _____

21. _____
22. _____
23. _____
24. _____
25. _____
26. _____
27. _____
28. _____
29. _____
30. _____
31. _____
32. _____
33. _____
34. _____
35. _____
36. _____
37. _____
38. _____
39. _____
40. _____

Name _____

Section _____

MINERAL IDENTIFICATION QUIZ #2

1. _____
2. _____
3. _____
4. _____
5. _____
6. _____
7. _____
8. _____
9. _____
10. _____
11. _____
12. _____
13. _____
14. _____
15. _____
16. _____
17. _____
18. _____
19. _____
20. _____
21. _____
22. _____
23. _____
24. _____
25. _____
26. _____
27. _____
28. _____
29. _____
30. _____
31. _____
32. _____
33. _____
34. _____
35. _____
36. _____
37. _____
38. _____
39. _____
40. _____

Rocks and Their Origins

2

INTRODUCTION

Rocks are familiar to all of us as the natural material exposed in areas of little or no vegetation cover, such as cliffs, roadcuts, quarries and outcrops. Rocks vary in color and texture, and often appear layered or fractured. In this chapter, we will examine in detail various types of rocks, the processes which produce them, and their important physical characteristics, with a view to field identification and an understanding of their genesis. Students should visit nearby localities to examine rock bodies *in situ*, as well as dealing with hand samples in the laboratory. The study of rock origins and classification is called **petrology**.

DEFINITION

Rock is any coherent, naturally occurring substance, usually composed of minerals.

This broad definition includes many different kinds of earth materials. Some rocks, such as rock salt, are composed of only one mineral. Granite *must* contain quartz and feldspar—and almost always includes accessory minerals as well. The rock conglomerate is literally a conglomeration of rounded fragments of other rocks. Coal is composed of compressed plant remains. Volcanic glass, called obsidian, is also a rock although it cooled so quickly that no minerals were able to form. Solid substances which make up the other planets must also be considered rock. Even glacial ice is a rock.

The definition does exclude many earth materials, some of which are common and familiar. Beach sand and river gravel are not rock because they are loose and uncemented. Lava is a liquid, and cannot hold its own shape, so it is not considered to be a rock. And concrete, which can look and perform very much like rock, does not qualify because it is man-made rather than natural.

Islands in the Pacific, such as Hawaii, are made of rock formed during volcanic eruptions. The reefs which grow quietly on their flanks, built by countless coral animals, are also rock. Soluble salt which forms a surface crust as at Badwater, in Death Valley, is as much a rock as the body which collided with the Earth to form Meteor Crater in Arizona.

Because the spectrum of rock types and rock-forming processes is so large, geologists have found it useful to subdivide them into the three major **families** discussed below.

CLASSIFICATION

Igneous rocks are produced when melted rock material, called magma, solidifies. Magma is generated by heat within the Earth. Some suspended crystals and dissolved gases often occur in the molten rock material. Lava is magma which flows out onto the Earth's surface. The term igneous comes from the Latin word *ignis*, meaning fire, and implies that heat is needed to create magma.

Sedimentary rocks form at the Earth's surface. Some sedimentary rocks are consolidated layered collections of particles washed, scraped, or blown off the land and deposited in low places such as lake and river bottoms, and the ocean floor. Other sedimentary rocks are created by the direct precipitation of chemicals from solution, often assisted by biological processes. The term sedimentary is from the Latin word *sedimentum*, meaning that which has settled out.

Metamorphic rocks result from the application of heat, pressure or chemicals on a rock in sufficient intensity to change the rock in some detectable way. Metamorphic is derived from the Greek words *meta*, meaning change, and *morphé*, which means form or shape.

Included in this unit are charts which describe the physical characteristics of the more common rocks in each family, along with some information about their mode of origin and the rock bodies they form. The charts emphasize the *differences* between rock families, and individual rock **types**. In nature, however, a complete spectrum of transitional examples links any two rock families, or any two types within a family. For this reason assigning an appropriate name to a given rock requires careful analysis of the specimen. All rocks are the result of some process acting on previously formed rock material. At the end of this unit we will examine the relation between rock classes, the cyclic nature of rock-forming processes, and the dynamics of the natural setting where the cycle takes place.

IGNEOUS ROCKS

Igneous rocks are the most abundant materials in the Earth's crust. They comprise about 80 percent of the volume of continents, and over 90 percent of the oceanic crust. In North America, an old largely igneous terrain called the Canadian Shield crops out over five million km^2 (three million square miles). About one-half million km^2 of the Pacific Northwest have been covered by the igneous rocks of the Columbia Plateau. Most large continental mountain ranges contain considerable igneous material; and some, like the Sierra Nevada and the Cascade Range, are made up almost entirely of rocks from this family. Additionally, nearly all of the oceanic mountains and islands of the world are igneous.

The formation of igneous rocks may be gentle and slow, or violent and swift, but it always involves the solidification of magma.

Magma and Igneous Rock Formation

The production of magma, or melted rock, requires considerable amounts of heat. Geologists believe that this heat has two origins: heat from the decay of radioactive elements within the Earth, and heat from friction generated as enormous blocks of the lithosphere collide or grind past one another. The mineralogy of an igneous rock will reflect the chemistry and cooling history of the magma from which it crystallized.

Minerals on the Igneous Rock Classification Chart (Table 2-1) crystallize in a definite order as the temperature drops:

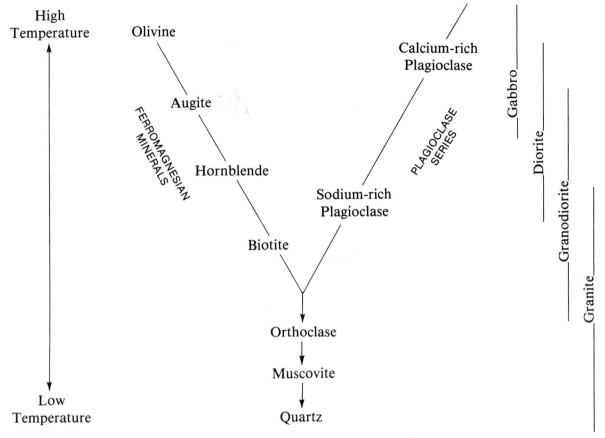

Figure 2-1. Crystallization Scheme

This crystallization scheme, described by N.L. Bowen in 1913, carries implications for the appearance and stability of the various igneous rock types. For instance, quartz in granite will have irregular grain boundaries because it forms last, and is forced to fill in between earlier developed crystals of feldspar and mica. Olivine grains in gabbro may have a corroded appearance if their outer surfaces were cannibalized to contribute iron and magnesium to later forming ferromagnesian species. Rocks such as granite which contain the later forming minerals are more stable when exposed at the Earth's surface because their temperature of formation more closely approximates surface conditions than the earlier formed types.

The distribution of igneous rock types is determined by the type of magma available at depth. Felsic rocks can only be found within continental areas; ocean basin and mantle igneous rocks are mafic. Rocks of intermediate composition are most commonly located at continental margins where mixing of continental and oceanic materials occurs.

Igneous Rock Types

Igneous rocks are classified according to the two aspects of their genesis described above—where the parent magma originates, and where it solidifies. Table 2-1 on page 33 incorporates these two fundamental parameters in designating common types of igneous rocks. Felsic rocks originating in the continental crust are at the extreme left, and rock types become progressively darker and more dense (mafic) toward the right approaching pure uncontaminated mantle-derived material.

The vertical dimension of the chart groups rocks on the basis of texture or crystal size. Rocks in the two bottom horizontal rows are **plutonic** (intrusive), formed when the magma solidifies below the Earth's surface to produce mineral crystals in the visible size range. The relatively coarse texture of these rocks indicates that the rocks cooled slowly in a thermally insulated environment. This would allow sufficient time for larger crystals to form. **Pegmatites**, which may have crystals up to several meters long, result from crystal growth in a very fluid, gas-rich environment in which large crystals organize fairly rapidly.

The **volcanic** (extrusive) rocks solidify quickly on the Earth's surface, and the crystals are correspondingly small. Obsidian and its frothy equivalent, pumice, have no crystals whatsoever. **Porphyries** have a few visible crystals set in a fine-grained groundmass, and demonstrate initial slow cooling followed by rapid solidification. This happens when magma begins to crystallize slowly at depth, and then shoots to the surface where remaining minerals form quickly. The larger (visible) crystals in porphyries are called **phenocrysts**. Common phenocrysts vary in composition with rock type as noted below:

Rhyolite porphyry: quartz, muscovite
Dacite porphyry: biotite
Andesite porphyry: hornblende, augite
Basalt porphyry: olivine, plagioclase feldspar, augite

A thoughtful use of the igneous rock chart should enable the student to do three things:

a. Identify (name) rock specimens provided in lab or observed in the field. For example, a specimen containing hornblende phenocrysts in a fine-grained gray background is most likely andesite porphyry.

b. Form a mental picture of rocks listed on the chart, considering minerals learned in Chapter 1, and crystal size indicated. For example, a peridotite will have abundant visible crystals of olivine along with other dark crystals in the visible size range.

c. Make a statement about the source of magma and cooling history of a specific rock. For example, obsidian is a melted and resolidified portion of the continental crust which formed quickly at the Earth's surface.

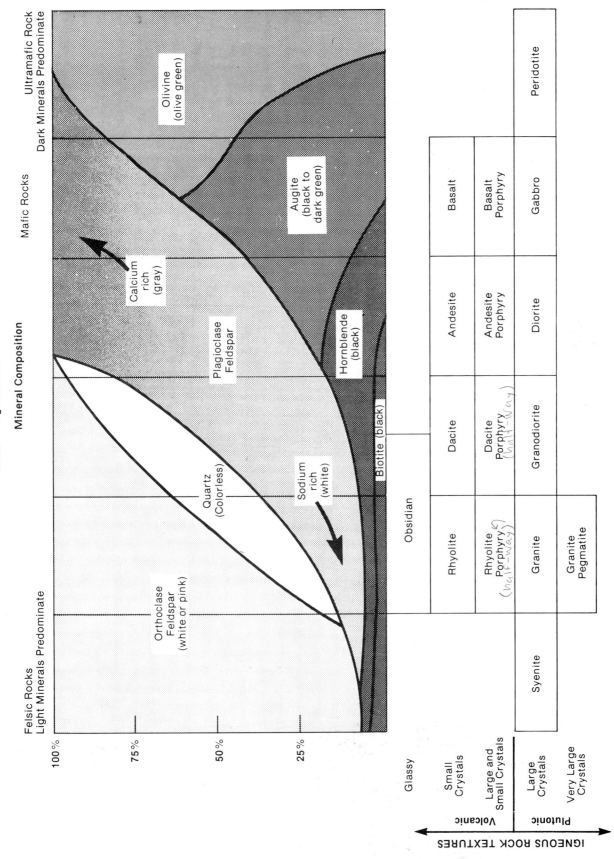

Table 2-1. Igneous Rock Classification

34 Exploring Geology

Volcanic Rocks   **Plutonic Rocks**

 ◄ Chemical Equivalents ►

Rhyolite tuff breccia. (Photo by Gary Jacobson.)   Granite. (Photo by Gary Jacobson.)

 ◄ Chemical Equivalents ►

Andesite porphyry. (Photo by Gary Jacobson.)   Diorite. (Photo by Gary Jacobson.)

 ◄ Chemical Equivalents ►

Olivine basalt porphry. (Photo by Gary Jacobson.)   Augite plagioclase gabbro. (Photo by Gary Jacobson.)

Photo 2-1. Igneous Rock Types

Plutonic Rock Bodies

Plutonic rocks occur as several types of intrusive bodies, categorized on the basis of their size, shape, and geometric relationship to rock present in the area prior to igneous invasion (country rock). The most common igneous bodies are:

Batholith: large (>100 km² exposed surface area) irregular shaped plutons with no known bottom.
Stock: small batholith (<100 km² exposed area).
Dike: tabular (sheet-like) pluton which cuts across the structure of country rock.
Sill: tabular pluton which is mostly parallel to the structure of country rock.
Laccolith: sill with a domed upper surface which has deformed the overlying rock.

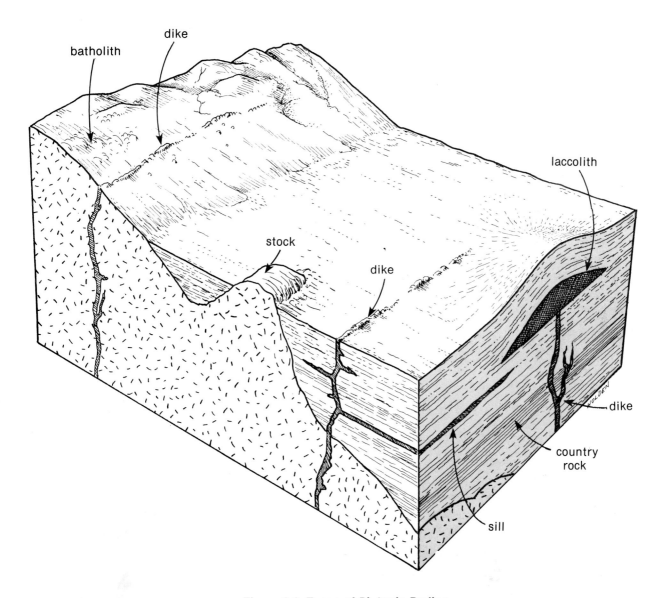

Figure 2-2. Types of Plutonic Bodies

36 Exploring Geology

A diabase sill has intruded between older sedimentary layers in the northern Rocky Mountains. (Photo by Shannon O'Dunn.)

This boulder of Colorado pegmatite shows large (.5m) crystals of muscovite and orthoclase. (Photo by Shannon O'Dunn.)

Massive columnar joints (columns to +2m diameter) in exposed laccolith, Devil's Tower, WY. (Photo by Evans Winner.)

Resistant dike etched out of country rock, Colorado Rockies. (Photo by Shannon O'Dunn.)

Expansion fractures (joints) in granodiorite, Southern California Batholith. (Photo by Shannon O'Dunn.)

Undigested country rock remnants (xenoliths) in granite, Southern California Batholith. (Photo by Shannon O'Dunn.)

Photo 2-2. Plutonic Rock Types and Bodies

Volcanic Landforms

Volcanic activity has created many famous mountains and dramatic terrains (Table 2-2). Most of these have developed where slabs of the Earth's rigid outer shell (lithosphere) are pulling apart, colliding, or are moving slowly over localized hot spots in the mantle. Thus, we find extrusive igneous processes located in well-defined zones, like the Pacific Ring of Fire and the Hawaiian Island chain.

Characteristics of various volcanic landforms are often determined by the type of magma extruded. Felsic magmas with considerable water content, generated where continental lithosphere melts, are associated with explosive volcanic events and the production of huge volumes of **pyroclastic** debris. Pyroclastic (fire-broken) is the term for magmatic materials which have been explosively ejected from a vent, and are partly or totally solidified before hitting the ground. Because larger amounts of ejected material statistically collect nearer the vent, tall cone-shaped *composite* volcanos are formed as the pyroclastics alternate with lava flows.

Dry, low viscosity mafic magmas originating in the upper mantle usually emerge quietly, and flow a long distance before solidifying. Flows are relatively thin, but areally extensive, and pyroclastic deposits are rare. Resulting landforms have a subdued profile. The cooling effect of sea water retards the spread of oceanic lava flows. Basalt lavas on land will first fill in topographic irregularities, and then continue to build great thicknesses of nearly horizontal flows called plateau basalts or *fissure flows*.

The classic shield shape of Mauna Loa rises above more frequently active Kilauea on Hawaii. (Photo by Shannon O'Dunn.)

Columnar jointing in basalt lava flow, the Devil's Postpile, California. (Photo by Roland Brady.)

Amboy Crater, a very young (2,000 to 6,000 yrs.) basalt cinder cone, Mojave Desert, California. (Photo by Chuck Monds.)

A wisp of smoke may be seen emerging from the summit of this active New Zealand composite cone. (Photo by Shannon O'Dunn.)

Photo 2-3. Volcanic Landforms

Table 2-2. Volcanic Landforms

	Cinder Cone	Composite Volcano (*Stratovolcano*)	Shield Volcano	Fissure Flows
Base width	←— 1 Km —→	←— 10 Km —→	←— 100 Km —→	+ 100 Km
Angle of Slope	30°–40°	5° Base 30° Summit	10° Base 2° Summit	± 0°
Tectonic Setting	Continental, and above sea level in oceans.	Colliding plates; coastal setting	Separating plates, hot spots; oceanic setting	Separating plates; continental setting
Rock Type and Texture	*Rhyolite, andesite,* or *basalt*. Built of pyroclastic* materials. Usually associated with flows.	Average composition is equal to *andesite*. Built of flows and pyroclastics interlayered.	*Basalt*. Built of flow accumulations with minor pyroclastics.	*Basalt*. Successive flows, fed by multiple dikes. Rare pyroclastics.
Activity	Usually one episode of explosive activity lasting for a few hours, days or weeks.	Repeated events every few hundred years from one to several million years. Some explosive events, some quiet. Occasional fiery clouds.	Events every few years for several million years. Relatively quiet eruptions.	Events every few thousand years over millions of years. Quiet eruptions.
Examples	Sunset Crater, AZ; Amboy Crater, CA; Paricutin Volcano, Mexico.	All large Pacific Ring of Fire (the Cascades, Andes, Japan, the Philippines, etc.) and Mediterranean volcanos.	All deep ocean basin volcanos as Hawaii, Iceland, the Canary Islands, Galapagos.	The Columbia Plateau of Washington-Oregon-Idaho; other notable localities in India and Africa.

*Pyroclastic: Magmatic materials which are thrown out of a volcano, solidify in the air, and settle to the ground as ash, lapilli, bombs. Consolidated pyroclastic debris forms a rock called **tuff**.

SEDIMENTARY ROCKS

Sedimentary rocks form a thin, discontinuous blanket over much of the Earth's surface above sea level, and constitute about two percent of the volume of the lithosphere. The greatest thicknesses of sedimentary rock exist in downwarped basins, like the Great Valley of California and the Gulf Coast region of the United States, where more than ten kilometers of these materials have been penetrated in the search for petroleum.

Sedimentary rocks are formed by the consolidation of sediment: loose materials delivered to the depositional sites by water, wind, glaciers, and landslides. They may also be created by the precipitation of calcium carbonate, silica, salts, and other materials from solution.

Sedimentary Environments

Sedimentary rocks form on continental platforms and in the oceans. Figure 2-3 shows various depositional sites within the continental and marine environments.

In order to understand the present day distribution of sedimentary rocks we must keep in mind that in the geologic past, large areas of North America (and the other continents) have been covered with shallow seas, experienced equatorial conditions, or supported extensive glaciers. For this reason there are shallow marine deposits exposed in the Grand Canyon, coral reefs and evaporated salt beds in the Great Lakes region, and glacial deposits over much of the northern Midwest.

Sedimentary Rock Types

Sedimentary rocks are subdivided into two classes. **Detrital** (clastic) rocks are those composed of fragments (clasts) of pre-existing rocks cemented or compressed into a coherent mass. **Chemical** and **biochemical** (non-clastic) rocks are produced when dissolved constituents in terrestrial or marine waters are precipitated by evaporation, chemical change, or biological activity.

The charts on pages 42-43 describe the most common representatives of these two rock groups. Detrital rocks are primarily distinguished on the basis of the largest common grain size. Chemical and Biochemical rocks tend to be more or less pure chemically, and thus can be identified in much the same way as their mineral components.

Distinctive Features

Many distinctive features of sedimentary rocks visible in hand specimens or outcrops give the geologist clues to the rock's origin and environment of deposition. A list of some of the more common features is given below, and several are shown in the photos on pages 41 and 44.

  a. **Particle Size:** The size of the particles is an indication of the energy of the transporting medium. For example, swift streams carry cobbles; wind and waves transport sand grains; and clay particles often float far out to sea on gentle currents.
  b. **Stratification:** Strata, beds, or layers are formed by repeated depositional events, or by a change in the material supplied to the depositional site. Stratification is the most common feature of sedimentary rocks.
  c. **Cross-Bedding:** A form of stratification in which layers between major strata are inclined to the horizontal. These form where depositional surfaces are at an angle, such as on the face of a sand dune or delta, or where sediment is delivered from different directions.
  d. **Concretions:** A localized concentration of cementing material, these are usually resistant to erosion and may stand out from the rock surface as lumps or bulges.
  e. **Jointing:** A regular pattern of cracks usually perpendicular to bedding planes and caused by breakage due to the weight of overlying rocks.
  f. **Ripple Marks:** Small waves or ripples formed by the movement of water or wind over the surface of the sediment prior to solidification.

40 Exploring Geology

g. **Fossils:** Any evidence of past life preserved in the rock. May be bone or shell fragments, footprints, leaf imprints, or organic materials replaced by silica or other chemicals.

h. **Color:** Most colors, including red, brown, ochre, green, and purple, are due to various iron compounds. Black is commonly caused by organic material, and white usually indicates some salt, clay (i.e. kaolinite), or silica.

i. **Cementing Agents:** Calcium carbonate, silica, and iron oxides are the most common binding agents in clastic sedimentary rocks.

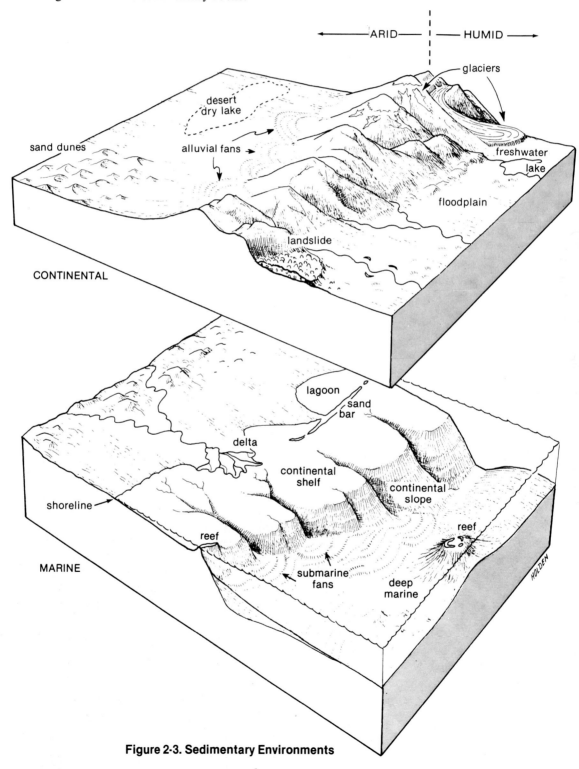

Figure 2-3. Sedimentary Environments

Pebble conglomerate with sandstone matrix. (Photo by Gary Jacobson.)

Pebble to boulder conglomerate exposed in a sea cliff, Pacific Beach, California. (Photo by Gary Jacobson.)

Cross-bedded Navajo sandstone at Zion National Park. (Photo by Shannon O'Dunn.)

Concretions in sandstone cemented with calcium carbonate, La Jolla, California. (Photo by Shannon O'Dunn.)

Fossil fish (.3m) in diatomaceous shale from Lompoc, California. (Photo by Gary Jacobson.)

Well-developed bedding in shale, near Monterey, California. (Photo by Roland Brady.)

Photo 2-4. Detrital Sedimentary Rocks

DETRITAL SEDIMENTARY ROCKS

Particle Size	ROCK NAME	DISTINCTIVE FEATURES	DEPOSITIONAL ORIGIN
Granular to Boulder-size Particles 2 mm and larger	Conglomerate	Rounded grains up to boulder size set in finer material such as sandstone or siltstone.	High energy environments: stream and river beds, submarine canyons.
	Breccia	Angular grains up to boulder size set in finer material such as sandstone or siltstone.	Little or no transport of clasts. Deposited by flash floods/mudflows, landsliding, and limestone cavern collapse.
	Tillite	Occasional large angular and/or rounded clasts set in silt or clay. Massive structure.	Glaciers. Widespread deposits formed by continental ice sheets.
Sand-size Particles 1/16 mm - 2 mm	Quartz sandstone (cleaner)	Quartz grains may be angular or rounded, clear or frosted. May be stained with iron oxide. Will scratch glass. May have some feldspar.	Sand dunes, beach faces, offshore bars. Settings where durable quartz is winnowed from less stable or less resistant grains.
	Arkose (Ortho. rich)	Contains 25% or more orthoclase with the quartz. May have micas. Grains usually angular and coarse. May resemble the granite from which it came.	Typically, the weathered debris of granite deposited on local alluvial fans or floodplains.
	Graywacke	Gray or greenish-gray, dense, fine-grained sandstone. Quartz rare; feldspars and rock fragments common. Usually has angular sand-size particles in dark silt or clay matrix.	Rapid deposition in offshore marine locales by submarine slumping or underwater mudflows, usually in tectonically active zones.
Clay to Silt-size Particles 1/256 mm - 1/16 mm	Siltstone	Fine-grained rock with slightly gritty feel. Will separate along bedding planes with difficulty.	Moderately high energy aqueous environments: rivers, nearshore marine.
	Shale	Smooth feel due to very small (clay-size) particles. Splits easily along closely spaced bedding planes. *grey-ocean basin; black-carbonaceous swamps (anaerobic environ.); red-mean land*	Low energy aqueous environments: continental shelves, lagoons, deep marine, lakes.

CHEMICAL AND BIOCHEMICAL SEDIMENTARY ROCKS

	ROCK NAME	DISTINCTIVE FEATURES	DEPOSITIONAL ORIGIN
Reacts with HCl [the carbonates]	Limestone	Reacts vigorously with HCl. Usually massive, light colored. May have conchoidal fracture, fossils, small oval or spherical structures.	Organic or inorganic precipitation in marine or fresh water: reefs, deep sea floor, caverns, hot springs.
	Dolomite	Reacts sluggishly with HCl. Usually massive, light colored. Fossils rare.	Shallow to deep marine. Inorganic.
	Chalk	Reacts vigorously with HCl. Usually porous and easily broken. A compacted powder of microscopic calcareous shells.	Mostly shallow marine. Organic.
	Coquina	Shell fragments cemented with calcite. "Shell breccia."	Shallow marine settings rich in shell-building organisms.
No reaction with HCl — Light color — Will not scratch copper coin	Rock salt	Rock form of halite. Usually a coarse crystalline aggregate. Colorless, pale orange, other colors.	Sites of evaporation of natural waters: coastal marine, desert 'dry' lakes.
	Rock gypsum	Scratches with fingernail. Usually in compact granular masses. May exhibit fibrous texture.	Same as halite.
	Diatomite	White, low density. Finely bedded. A compacted powder of microscopic siliceous diatom shells.	Fresh water lakes. Marine settings rich in dissolved silica. Organic.
No reaction with HCl — Light color — Will scratch copper coin	Chert	Massive. Conchoidal fracture. Broken surfaces resemble unglazed porcelain.	Deep ocean oozes of microscopic animal shells. Altered diatomite. Inorganic precipitation of silica from sea water.
No reaction with HCl — Dark color	Coal	Dark brown to black, low density. Often banded. May display plant remains.	Decomposed and altered remains of terrestrial plants laid down in bogs, swamps, and estuaries.
	Flint	Gray to black variety of chert. Often has a white coating of chalk.	Same as chert.

44 Exploring Geology

Fossiliferous limestone. (Photo by Gary Jacobson.)

Fine-grained limestone with dendrites. (Photo by Gary Jacobson.)

Chert. (Photo by Gary Jacobson.)

Agatized tree trunk, Petrified Forest, Arizona. (Photo by Roland Brady.)

Selenite gypsum. (Photo by Gary Jacobson.)

Halite crystals on forked twig, Salton Sea, California. (Photo by Gary Jacobson.)

Photo 2-5. Chemical and Biochemical Sedimentary Rocks

METAMORPHIC ROCKS

Formation of Metamorphic Rock

Rocks form in equilibrium with their environment, and are therefore most stable under the conditions which prevailed at the time of their formation. Because the lithosphere is a dynamic system, rocks are often transported to a new setting where pressure, temperature, and chemical conditions differ significantly from those where they were formed. If these new conditions produce a significant, detectable change in the texture or mineralogy of the original rock, it is considered metamorphosed.

When metamorphic conditions are relatively mild or short-lived, **low-grade** metamorphic rocks result. Traces of original features such as stratification, fossils, and gas bubbles may still be detectable. **High-grade** (more severe) conditions generally tend to promote the growth of new mineral species at the expense of original ones, and obliterate pre-existing textures and structures.

Heat and leftover fluids escaping from a cooling batholith will alter surrounding country rock to create **contact metamorphic** aureoles. The most severe changes take place adjacent to the igneous contact. Metamorphic effects, which are largely temperature controlled, fade rapidly as distance from the pluton increases. Chemical changes are most profound in situations where pluton chemistry differs radically from that of the country rock.

Heat and pressure generated where lithospheric plates intersect create intense metamorphic changes over wide zones as the plate boundary is approached. Metamorphic rock types thus produced vary with the nature of the orginal rock. A hypothetical section through a progressive **regional metamorphic** sequence with various sedimentary rocks and lavas is represented in the sketch below.

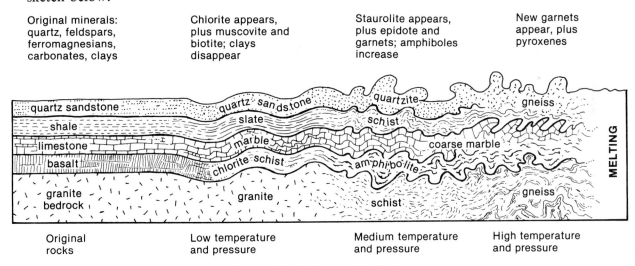

Figure 2-3. Metamorphic Facies

Metamorphic Rock Classification

Classification of the metamorphic rocks is based on their texture and mineralogy. (See p. 46.)

Non-foliated or **massive** rocks are generally simple in mineralogy, and contain one dominant chemical constituent. The previous form (parent rock) may be assigned with considerable confidence because of this simplicity.

Foliated rocks are characterized by parallel platy or elongated minerals which aligned in response to direct stress during metamorphism. The parent rock may be obscure. However, the size of the mineral crystals is thought to be an indication of the severity and/or duration of application of heat and pressure. Foliated rocks trend from low-grade metamorphic slate to high-grade metamorphic gneiss (Fig. 2-3 above).

METAMORPHIC ROCK CLASSIFICATION

	Color	Rock Name	Distinctive Features	Typical Parent Rock
Non-Foliated	Light Color	Marble	Reacts with hydrochloric acid. Color streaks or blotches may be present. Look for calcite rhombohedrons if coarse. Rare "ghost fossils."	Limestone
	Light Color	Quartzite	Fused quartz grains will fracture across original grain boundaries. May have a sugary texture, but smoother than sandstone.	Sandstone
	Green	Serpentinite	Lime green to dark green or black, heavy and dense. Commonly has slickensided surfaces.	Mafic or Ultramafic
	Green	Eclogite	Granular dark green rock studded with red garnets; may show weak foliation.	Ultramafic
	Dark Gray to Black	Hornfels	Dense, fine-grained rock with conchoidal fracture.	Any fine-grained rock
	Dark Gray to Black	Anthracite	Shiny, low density black rock; may have conchoidal fracture and display parting or banding. *Highest B.T.U.*	Bituminous coal

	Crystal Size	Rock Name	Distinctive Features	Parent Rock
Foliated	Microscopic Crystals	Slate	Dull to shiny; splits into thin slabs. Harder than shale. Commonly dark gray, brown, red, green.	Siltstone, shale, silicic volcanic rocks
	Microscopic Crystals	Phyllite	Nearly invisible mica crystals give this rock a satiny sheen on foliation surfaces. Often gray or gray-green.	Siltstone, shale, slate
	Large Crystals	Schist	Visible aligned platy or elongate minerals cause foliation. Quartz, feldspar or garnets common.	Volcanic rocks, shale, slate, phyllite
	Large Crystals	Amphibolite	Dark, heavy rock with aligned hornblende crystals and accessory feldspar.	Mafic igneous rocks, graywacke, carbonate-rich shale
	Large Crystals	Gneiss	A coarse-grained rock with banded appearance due to mineral segregation.	Silicic igneous rocks, arkose, siltstone, or shale

Quartzite. (Photo by Gary Jacobson.)

Serpentinite with prominent slickensides. (Photo by Gary Jacobson.)

Anthracite coal. (Photo by Gary Jacobson.)

Slate. (Photo by Gary Jacobson.)

Biotite schist. (Photo by Gary Jacobson.)

Gneiss showing distorted foliation and small cross-cutting dike, Southern California Batholith. (Photo by Allen Bassett.)

Photo 2-6. Metamorphic Rock Types

ROCK FAMILIES AND THE DYNAMIC LITHOSPHERE

The concept that all rocks are formed from pre-existing rocks in a continuing cycle is a fundamental principle in geology. We have seen in this chapter that members of the igneous, sedimentary, and metamorphic rock families are discrete phases produced by outer Earth processes acting on pre-existing material.

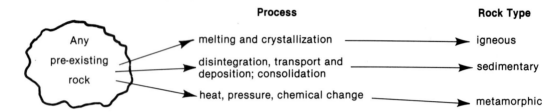

The implications of this concept are profound:

> 1. All earth materials, excepting meteorites, are original constituents of this planet, and were present at the time the Earth was completely formed about 4.5 billion years ago.
> 2. The three-family rock classification scheme categorizes rock materials on the basis of the most recent major events or processes they have experienced.
> 3. These processes require energy—chemical, thermal, and gravitational. Planet Earth is a dynamic place where chemical and gravitational gradients still exist, and thermal energy is available.

Transformation of earth materials from one rock family to another usually occurs in tectonically active zones where moving lithospheric plates are interacting. Chapter Eight will discuss the nature of these interactions in detail. For the present, we may consider the following generalizations:

> **Volcanic igneous rocks** are created in three tectonic settings: where plates pull apart, releasing melted mantle in fissure flows and shield volcanos; where plates collide and melt rocks through friction; and over local mantle hot spots. **Plutonic rocks** are mostly formed at plate collision boundaries, where some friction-melted magma is trapped in batholiths below the surface.
>
> **Metamorphic rocks** in quantity are produced by pressure and frictional heat generated where plates collide. This plate collision provides energy for regional metamorphic transformations. Halos of metamorphosed country rock altered by heat and fluids released from an adjacent batholith often surround these massive igneous bodies.
>
> **Sedimentary rock** deposits of considerable thickness require vertical crustal motions to create uplifted source areas and catchment basins. Colliding plates push up parts of the lithosphere to form sediment-shedding mountain ranges. Basins may be created where plates sliding past one another cause deep tensional tears in the lithosphere, or where separating continental slabs open up elongate troughs called rift valleys. Because the ultimate requirement for deposition is a gravity sink, you will note that many sedimentary sequences collect in low quiet areas, such as deep ocean bottoms, away from plate boundaries.

Rock masses maintain their identity as specific representatives of one of the rock families for varying lengths of time, depending upon their tectonic environment. Basalt flows produced at separating plate boundaries may have a relatively short ride of a few million years before being remelted at a collision site. In contrast, some sedimentary sequences located in plate interiors have survived unaltered for more than three billion years. Generally speaking, most rock masses we observe today will be transformed into new rock types in the geologic future. Also, the closer the rock mass is to an active tectonic zone, the greater the certainly of transformation.

Rocks 49

Table 2-3. Rock Symbols

The symbols below are the conventional representations for rock types described in this chapter. They will be used on maps, cross-sections and other diagrams throughout the remainder of the book. Additional symbols/colors representing specific rock types will be introduced for individual exercises where appropriate.

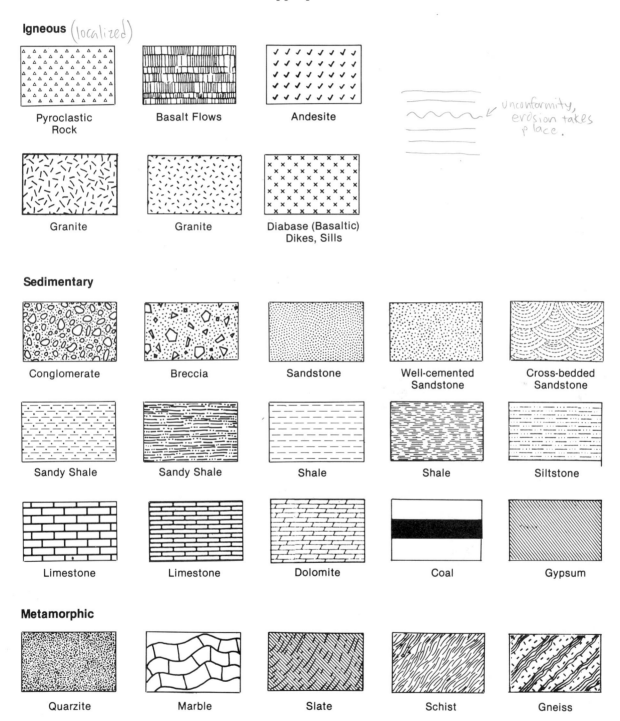

Annotations: "Igneous (localized)"; near wavy lines: "← unconformity, erosion takes place."

50 Exploring Geology

Rock Sample Identification

Specimen Number	Rock Family	Physical Characteristics	Rock Name
	Igneous →	Large crystals, small crystals, porphyry; minerals present; bubbles; pyroclastic texture; flow banding; etc.	
	Sedimentary →	Grain size if clastic; mineralogy if chemical or biochemical; bedding, fossils; internal structure; etc.	
	Metamorphic →	Non-foliated: color, reaction w/ HCl, density, etc., Foliated: crystal size, composition, banding, etc.	

Rocks 51

Chapter 2: Rocks Name_____

Section _____

REVIEW QUESTIONS

1. What are the *differences* between rocks and minerals?

2. The plutonic rock with the highest mafic mineral content is _____.

3. What two igneous groups are distinguished by *where* the magma cooled?

   _____   _____

4. What is the most universal feature of sedimentary rocks? _____

5. Fractures in rocks caused by expansion or pressure are called _____.

6. Fossils are almost exclusively associated with the _____ rocks.

7. The sedimentary rock type deposited by glacial ice is _____.

8. What do dikes and sills have in common? _____

   _____

   How do they differ? _____

   _____

10. What element is usually responsible for each of the following sedimentary rock colors?

    a. Brown   _____

    b. Red   _____

    c. Black   _____

11. Name two common cementing agents in sedimentary rocks:

    _____   _____

12. The Hawaiian Islands are formed by _____ type volcanos of _____ rock composition.

13. What are the metamorphic equivalents of the following rocks?

    a. Limestone = _____

    b. Basalt   = _____

    c. Sandstone = _____

    d. Slate   = _____

    e. Shale   = _____

52 Exploring Geology

14. The two fundamental parameters used to classify the igneous rocks are _____ and _____.

15. The largest plutonic bodies are called _____.

16. What plutonic rock contains the most plagioclase feldspar? _____.

17. A quartz-free plutonic igneous rock is _____.

18. Peridotite magmas originate in the _____.

19. Composite volcanos are created where lithospheric plates _____.

Chapter 2: Rocks Name _____

Section _____

MORE CHALLENGING QUESTIONS

1. Give two examples of materials considered to be both rocks *and* minerals. _____

   _____

2. The volcanic rock formed by mixing continental and mantle material is _____.

3. Most biochemical sedimentary rocks are formed in the _____ environment.

4. What are the names for the following igneous rock textures?

   a. Very large crystals   _____

   b. Mixed visible and microscopic   _____

   c. No crystals at all   _____

5. What are the agents of metamorphism? _____

6. A porphyry containing calcium-rich plagioclase phenocrysts is most likely _____.

Name _____

Section _____

ROCK IDENTIFICATION QUIZ

1. _____
2. _____
3. _____
4. _____
5. _____
6. _____
7. _____
8. _____
9. _____
10. _____
11. _____
12. _____
13. _____
14. _____
15. _____
16. _____
17. _____
18. _____
19. _____
20. _____

21. _____
22. _____
23. _____
24. _____
25. _____
26. _____
27. _____
28. _____
29. _____
30. _____
31. _____
32. _____
33. _____
34. _____
35. _____
36. _____
37. _____
38. _____
39. _____
40. _____

Aerial Photographs 3

INTRODUCTION

In 1859 Oliver Wendell Holmes, Sr., American poet, writer, and prominent physician, published an article in the *Atlantic Monthly* describing a gadget called a stereoscope. In this device one could place picture cards containing two photographs of the same scene taken from slightly different camera postions. Viewed in the stereoscope, the two cards merged to create a single scene which appeared in three dimensions. This so delighted people that it became standard equipment for evening entertainment in the Victorian era.

This toy slowly evolved into a tool that could be used in planning cities, laying out highways, and in other endeavors that require a three-dimensional view of the land. During the mid-1930's, as advances in photography and in aviation were made, aerial photographs received much attention in highway development both in Europe and the United States.

During World War II aerial photography, used for locating military installations in enemy territory, became so valuable that a whole new field of scientific endeavor, **photogrammetry**, developed rapidly to exploit the invention of stereoscopic viewing of photographs. Ultimately it was realized that this constituted a superb method of simplifying the task of making topographic maps and aiding the geologist in making geologic maps. This topographic and geologic mapping stimulated a new industry and has greatly expedited the business of map making.

This chapter describes the making of aerial photographs, how they are used for field mapping and to make topographic maps, and the latest advances in various methods of producing and enhancing aerial photographic data.

MAKING AERIAL PHOTOGRAPHS

There are many potential variations in aerial photography: vertical and oblique views, black and white, true color, false color, and infrared. Flight altitudes vary from near ground level to orbiting satellites. For vertical aerial photos a camera mounted in the bottom of an airplane takes vertical photographs of the ground along carefully monitored flight lines. Each picture is customarily taken on a 9''x9'' negative which covers an area of 10 to 20 square miles, depending upon the flight altitude (see Fig. 3-1). Comparing aerial photos to maps of known scale is helpful in determining distances between points shown on the photo. The scale can also be calculated by dividing the focal length of the lens (in feet) by the altitude of the aircraft (in feet).

58 Exploring Geology

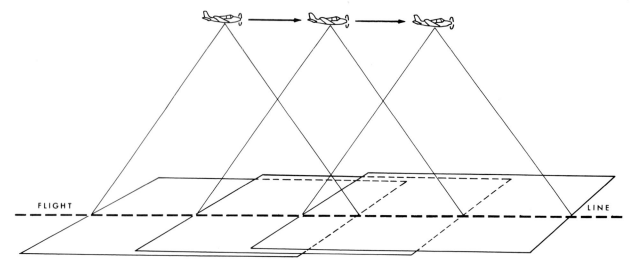

Figure 3-1. Consecutive Air Photographs with 60 Percent Overlap

If the photos are taken so that the same area on the ground appears in two consecutive photos taken from two different positions (usually with a 60 percent overlap), it is possible to view them stereoscopically and thus simulate the three-dimensional effect of being in the airplane. This effect is greatly exaggerated, so that the relief appears more extreme than in actuality. Mildly rolling terrain may look fairly rugged. This distortion is due to the fact that instead of gaining stereoscopic three dimensions from the 2½-inch separation of your eyes, you get a one-mile separation from the two positions of the airplane when the photos were taken.

Vertical photographs are often pieced together to create a **photomosaic** by carefully matching edges between adjacent photos, tearing them along irregular borders, and pasting them together to create a single image. These mosaics are then re-photographed and reduced in size to make pictures of entire quadrangles, states, or even whole countries (see Photo on page 96).

USING AERIAL PHOTOGRAPHS

Aerial photos are an important tool in geologic interpretation and field investigation. Slight variations in tones of gray may be due to differences in rock types and to changes in vegetation (which may also be geologically controlled). Among the approaches that can be used to interpret geology from these photos are the analyses of drainage patterns and the study of landforms, visible in their entirety only from the air. With these methods, geologists may create maps of great accuracy while spending little time in the field. This has been especially valuable in reconnaissance mapping of large previously unmapped areas, particularly those which are remote or physically inaccessible.

Overlapping stereo pairs are usually viewed through a stereoscope. Pocket stereoscopes are handy for both field work and laboratory use as they are portable and durable. Although only a narrow strip of the photos may be viewed, magnification enhances details of the surface and facilitates locating a point on the ground.

To use the pocket stereo, set the legs up and pull the lenses apart. *Be careful! If your nose is between the lenses when you push them back together to set the proper distance—you may break the bridge of your nose.* Set the lenses for your interpupillary distance, usually about 65 mm on the scale between the lenses, and place it over the square sets in Figure 3-2. You are seeing stereo correctly when you achieve an image of nine squares at various heights. Now try viewing Stereophoto Set 3-1 on page 61, placing each lens over the same point on each photo. You may need to rotate the viewer until the images merge.

Figure 3-2. Stereo squares

Large stereoscopes with a viewing system of lenses, prisms and mirrors are also used in the laboratory. With these instruments, two full aerial photos can fit side by side, and their entire overlap area can be viewed simultaneously without exaggeration (magnification). These stereoscopes are precision instruments which must be handled carefully; avoid contact with the mirror surfaces.

Most people without significant visual acuity differences can achieve stereo vision with the naked eye. To do this, hold Figure 3-2 up against your nose and look over the exact center of the page top at an object across the room. Then continue to focus in the distance as you slide the page up until the square sets are in front of your eyes. The images will be blurred, but should appear as one. Now slowly move the page out to about one foot in front of your eyes, keeping your focus at a distance and ignoring peripheral images. Try to bring the central square set into sharp focus to produce the same image you saw with the stereo viewer. Free-vision stereo may cause eye strain, and is not used routinely by most geologists.

AERIAL PHOTOGRAPHS AND TOPOGRAPHIC MAPS

Topographic maps, which show elevation changes on the ground by means of contour lines, were formerly sketched by hand in the field using a variety of standard surveying techniques for vertical and horizontal control. Today, most are compiled in the lab by photogrammetric methods utilizing aerial photographs and complex stereoscopic plotting instruments.

Elevation differences essential to constructing topographic maps are derived from vertical aerial photographs by using the phenomenon of **parallax**. When an object is viewed from two different positions, such as on overlapping photos, the object apparently shifts in position. This lateral shift, or parallactic displacement, is directly related to the height of the object, which can then be calculated mathematically.

Other kinds of information found on topographic maps, such as the extent of urban areas, placement of roads, and coastal alteration by building and dredging, are periodically updated by the United States Geological Survey using current photos of previously mapped areas.

ADVANCES IN AERIAL PHOTOGRAPHY

Tremendous advances in photographic imagery of the Earth's surface have been made in recent years, and some of the more notable ones are discussed here.

Vertical aerial photos in true color are becoming more generally available as technology continues to improve picture quality and reduce costs. These photos are particularly useful for geologic mapping, where various rock bodies must be distinguished and their boundaries plotted.

Infrared aerial photos are those in which color on the photos represents the relative heat of surface areas, rather than their visible color (*true color*) in reflected light. In general, colder areas show up in darker tones of blue, heat absorbing vegetation and other surfaces appear in reds and browns, and recently stripped or graded plots are light or white. Various types of vegetation (trees versus grass) are frequently apparent; and the coastal outline of lakes and the ocean, where colder water abuts the relatively warmer land, is starkly delineated.

Orthophotoquads are now being published at a scale of 1:24,000 by the U.S.G.S. These maps consist of contour lines superimposed on an actual photomosaic of the land, and thus combine the advantages of plain photos and topographic maps.

Landsat spacecraft, which are *satellites* orbiting at an altitude of 920 km (560 mi.) in a near-polar track, collect repetitive and multispectral data. This instrument is producing worldwide surface pictures, plus continually updated data on weather patterns and other rapidly changing phenomena. Other valuable images have been collected by high-altitude aircraft and by Skylab.

Side-Looking Airborne Radar (SLAR) is an electronic image-producing system in which radar beams are propogated perpendicular to the side of the aircraft. SLAR provides its own illumination in the form of microwave energy; thus images may be obtained day or night, and through cloud cover. Oblique SLAR views enhance subtle surface features and aid geologic interpretation.

Finally, information derived from aerial photographs of various kinds is now being used to make maps which show fine details of *land utilization*, *potential slope stability*, and *water and air pollution*. These will be invaluable to economic geologists, hydrologists, city planners, ecologists, and other experts in a variety of fields.

SELECTED STEREOPHOTO PAIRS

The following stereo photo pairs will introduce you to stereo viewing and some of the interpretive possibilities of photographic imagery in geology. Exercises which involve aerial photography will be found on page 67.

Aerial Photos 61

Stereophoto Set 3-1. Joint patterns in sedimentary rocks. Emery County, Utah. Joints are fractures along which no movement has occurred. They are common in hard and brittle rocks that resist folding. (National Archives Photo)

62 Exploring Geology

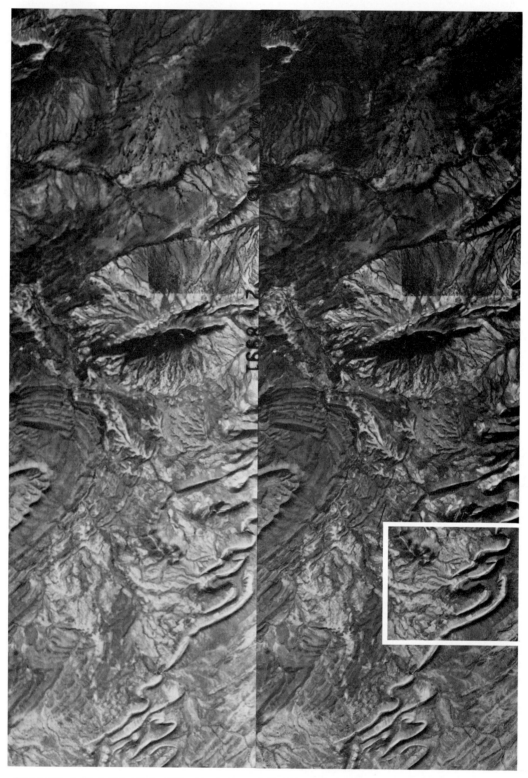

Stereophoto Set 3-2. High altitude photos of the Marathon Basin in West Texas showing complex folding and faulting of sedimentary layers. Inset marks the location of Photo 3-3. (NASA photo)

Photo 3-3. Low altitude oblique aerial photograph of a part of the Marathon Basin in west Texas. Oblique photographs provide a more three-dimensional view of the land and allow the interpretation of complex structures that are difficult to see on the ground. Compare with Stereophoto Set 3-2. (Photo and interpretation courtesy of Earl McBride.)

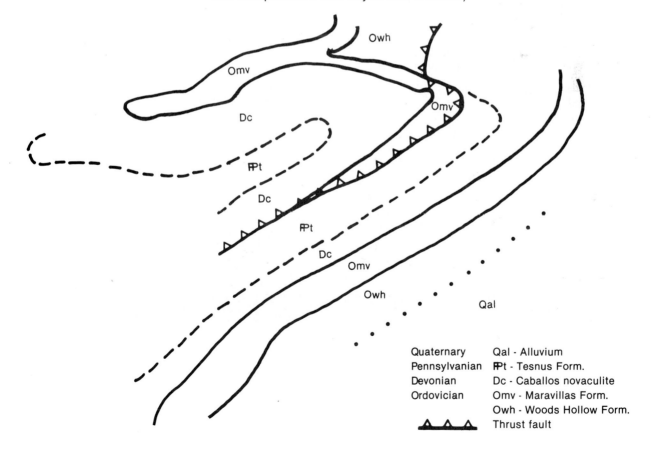

64 Exploring Geology

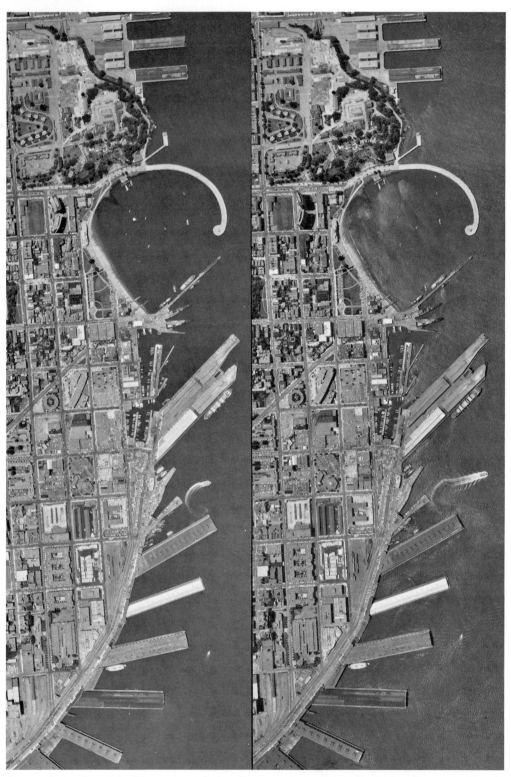

Stereophoto Set 3-3. Cityscape of the Embarcadero area, San Francisco, California. (California Dept. of Agriculture Photo)

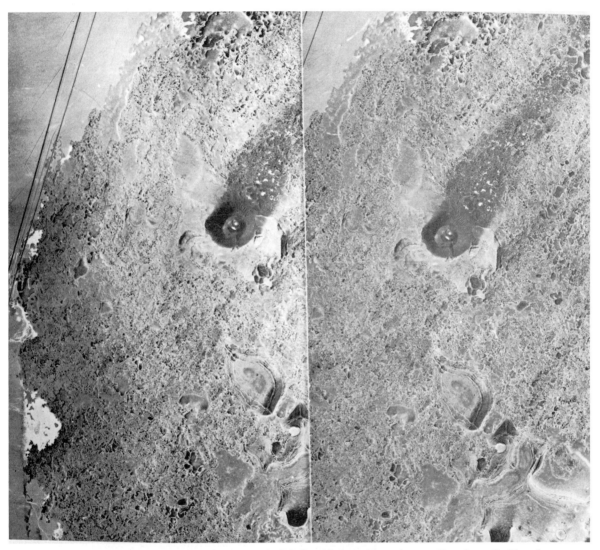

Stereophoto Set 3-4. Amboy Crater and vicinity, Mojave Desert, near Barstow, California. North is to the left in these photos.

66 Exploring Geology

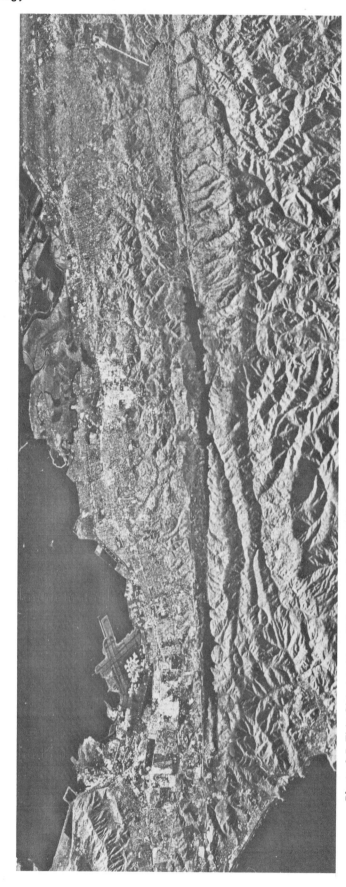

Photo 3-5. This Side-Looking Airborne Radar (SLAR) image is a view of a portion of the San Francisco Peninsula. The oblique view greatly enhances many of the surface features so that you can readily spot the San Andreas fault and see where it enters the Pacific Ocean on the left. The San Francisco airport is in the upper left part of the photo, while the Stanford Linear Accelerator (SLAC) is just above the fault traces on the right side. SLAC is two miles long. See Chapter 8 for further discussion.
(U.S.G.S. Photo)

Aerial Photos 67

Chapter 3: Aerial Photographs Name_____

Section _____

Aerial Photo Activities

Stereophoto Set 3-1.
   Sedimentary Rocks, UT

1. How many sets of joints can be distinguished? _____ If north is at the top, what are the compass directions of the joint sets? (measured from north) _____

2. How many different rock formations can be distinguished by the changes in the texture of the rocks?
_____

3. What kind of sedimentary rock is most likely to have joint systems? _____

   Why? _____
   _____

Stereophoto Set 3-2.
   Marathon Basin, TX

1. Is the circular structure at left a hill or valley? _____

2. What kinds of rocks would be most likely to form the thin, highly folded and faulted ridges on the right side of the photo? _____.

3. Is this an arid or humid zone? What is the most likely climate in this area? Explain. _____
_____

Stereophoto Set 3-3.
   Cityscape, CA

1. Which of the two photos, right or left, was taken first? _____ How can you tell? _____
_____

2. Why does glare on the water show in the photo on the right, but not in the left one? _____
_____

3. The natural feature with the highest elevation is located _____
_____

68 Exploring Geology

Stereophoto Set 3-4.
Amboy Crater, CA

1. From which direction was the wind coming when Amboy Crater erupted? _____

2. Which formed first: Amboy Crater or the lava flows surrounding it? _____

    How can you tell? _____

    _____

3. Lava from Amboy Crater poured out onto saturated salt beds of a desert "dry lake." What origin would you suggest for the depressions like the one about one inch west of the cinder cone? _____

    _____

Chapter 3: Aerial Photographs   Name_____

                                            Section _____

                            REVIEW QUESTIONS

1. Why are infrared photographs so useful in delineating shorelines?

2. When and why was the science of photogrammetry created?

3. A map which combines aerial photography with contour lines is called a(an)_____.

4. The underlying principle in the production of a stereo or three-dimensional image, either by the

   human eye or on overlapping photographs, is_____.

5. Compare the relative merits of the pocket stereoscope and the large, mirrored instrument.

6. What specific instrumentation is currently being used by the United States government to monitor fast-changing phenomena such as weather patterns and volcanic eruptions on a global scale?
   _____

7. Topographic maps, showing changing land elevations, used to be made by _____

   _____, but are now being produced with the aid of _____.

8. What kind of airborne imagery can see through cloud cover? _____

Stereophoto Set 3-4.
  Amboy Crater, CA

1. From which direction was the wind coming when Amboy Crater erupted? _____

2. Which formed first: Amboy Crater or the lava flows surrounding it? _____

    How can you tell? _____

    _____

3. Lava from Amboy Crater poured out onto saturated salt beds of a desert "dry lake." What origin would you suggest for the depressions like the one about one inch west of the cinder cone? _____

    _____

4

Fossils and Geologic Time

INTRODUCTION

The interpretation of geologic history has developed greatly since the 17th century, when the relative antiquity of a bone fragment was established by the fact that it would adhere to the tongue, and rocks were often classified as antediluvian or postdiluvian. The desire of students of Earth science to establish a chronology of terrestrial development and of the evolution of living organisms has transcended fundamental scientific misconceptions and religious bias to the point that the last half billion years, at least, are reasonably well understood. Considering that the estimated age of the Earth is 4½ billion years, this may not seem too impressive to the beginning student. However, as we have seen in Chapter 2, rock materials are constantly being recycled into *younger* forms, and fossils are destroyed. The record of most of the early Earth and of the evolution of life simply may not exist in any readable form anymore. Also, life on Earth prior to 600 million years ago was predominantly microscopic and without hard parts, and thus left a very poor fossil record. Therefore, in this chapter we will concentrate on the fossil record of the last one-ninth of the Earth's history, since it is in rocks of this age that the bulk of economic resources and recognizable fossils are found.

FOSSILS AND THEIR CLASSIFICATION

A fossil is defined as *any evidence of past life*, implying evidence of life which has not existed for some time. *Evidence* may be in many forms and need not involve actual remains; a frozen woolly mammoth, a prehistoric bone, petrified wood, and footprints are all considered valid fossils. Even worm burrows and the isolated gizzard-stones of dinosaurs qualify. The study of fossils is called **paleontology**.

The internationally used system of naming plants and animals, contemporary or fossil, was first presented by the Swedish botanist Carl von Linné in the 1758 edition of his book *Systema Naturae*. In the Linnaean classification plants and animals are grouped according to their inferred origin; i.e., their degree of similarity. The relationship is established by comparative anatomy and embryonic development. A paleontologist dealing with fossils rarely, if ever, has the opportunity to examine soft part anatomy or embryos, and so must rely heavily on hard part anatomy, such as shell fragments or bones. For this reason, the correct grouping of fossil organisms is necessarily somewhat speculative. Several known fossils defy attempts to place them in the most fundamental level of subdivision, the plant or animal kingdom. The complete Linnaean classification for three organisms is on page 72. Note that humans and dogs, being more closely related to one another than to a clam, share more categories:

72 Exploring Geology

Linnaean Classification	Modern Human	Domestic Dog	A Fossil Clam
Kingdom	Animalia	Animalia	Animalia
Phylum	Chordata	Chordata	Mollusca
Class	Mammalia	Mammalia	Pelecypoda
Order	Primates	Carnivora	Prionodesmacea
Family	Hominidae	Canidae	Pectinidae
Genus	*Homo*	*Canis*	*Pecten*
Species	*sapiens*	*familiaris*	*healeyi*
Individual	Carl von Linné	Spot	---

The scientific name of a living or fossil organism consists of the generic and specfic names (always italicized) along with the name of the person who first described it, and the year that description was published. The full scientific name for the fossil clam above is *Pecten healeyi* Arnold, 1906.

Applying the complete classification scheme above to a given fossil is often an intricate task requiring extensive knowledge of biological principles, terminology, and reference works. For our purposes, it will be sufficient to become familiar with the phyla which have left a significant fossil record. The brief descriptions below relate these phyla to familiar living animals. All these phyla span the time from Precambrian to Holocene (Table 4-1), but have had various representatives and different relative populations in each geologic period. Examples of each phylum are shown in Figures 4-2, 4-3, and 4-4.

Phylum Protozoa — One-celled aquatic animals. Fossil examples have external shells of calcium carbonate ($CaCO_3$) or silica (SiO_2). Marine and non-marine; mostly planktonic. Most common fossil phylum. (p. 79)

Phylum Porifera — The sponges. Silica or calcium carbonate skeletal elements appear in fossil record. Mostly marine.

Phylum Coelenterata — Animals having a hollow body with tentacles surrounding mouth. Includes sea anemones, jellyfish; common fossil representatives are the corals. (p. 79)

Phylum Brachiopoda — Marine bivalves, most with calcium carbonate shells. Look somewhat like clams, but each valve is bilaterally symmetrical. Most are attached to sea floor. (p. 79)

Phylum Mollusca — Most common macroscopic phylum. Includes clams, snails, octopus, and squid. Ancestors of the last having external shells are important as fossils. Shells usually calcareous. Mostly marine; also fresh water and terrestrial. (p. 80)

Phylum Arthropoda — Includes insects and crustaceans (crabs, lobsters, ostracods, shrimp, barnacles). Trilobite is the most important fossil. All have segmented chitin exoskeleton, and molt periodically. Marine, freshwater, terrestrial. Most populous phylum. (p. 81)

Phylum Echinodermata — All marine; commonly have 5-sided radial symmetry and calcareous plates forming exoskeleton. Star fish, sea urchins, sand dollars, brittle stars, and sea cucumbers are representatives; crinoids are common fossils. (p. 81)

Phylum Hemichordata — Includes the extinct *graptolites*, colonial pelagic marine organisms with slight development of a dorsal stiffening rod. Remains consist of small chitinous, sawblade-like films that resemble pencil marks on the rocks. (p. 81)

Phylum Chordata	Mobile animals with a dorsal nervous system and a notochord or vertebral column. Major subphylum is the Vertebrata which includes fish, reptiles, amphibians, birds, and mammals.
Phylum Conodonta	Extinct phosphatic microfossils that are very useful in correlation, but until recently of unknown zoological affinity. Recent discoveries indicate a possible relationship to the chordates.

BASIC PRINCIPLES OF STRATIGRAPHY

Geological analysis of a region or a continent is complete only when events, such as periods of sedimentation, erosion, structural deformation, and igneous activity, have been arranged in chronological order. Since World War II, absolute dates for various rock bodies and events have become reliable through the measurement of spontaneous radioactive decay. Treatment of absolute dating is beyond the scope of this manual; however, dates derived from this approach appear on the *Geologic Time Scale*, Table 4-1.

Historically, establishing a sequence of events (relative dating) has been accomplished through the study of rock body relationships and the fossils the rocks contain. Principles used in such studies are detailed below:

Superposition—This fundamental principle of sedimentary rock study is very simple. It states that in a sequence of sedimentary rocks piled up on the Earth's crust, the oldest stratum will be buried the deepest. Exceptions to this rule are uncommon, but do occur in tectonically active areas. Sections of strata which have been tilted more than 90 degrees are called **overturned**.

Original Horizontality—This principle states that most sedimentary materials settle under the influence of gravity and thus initially form horizontal layers. The implication here is that layers which are tilted or folded have been deformed subsequent to deposition.

Cross-cutting Relationships—This principle formalizes the common-sense observation that faults or intrusive bodies must be younger than the rocks they cut. Where a fine-grained igneous body is concordant with sedimentary layers, the distribution of baked contact zones must be examined in order to distinguish between a sill and a buried flow.

Faunal Succession—Use of fossils as an indication of geologic age depends upon two assumptions of organic evolution. The first assumption is, that organisms change over time and that a specific body plan is never repeated. This is another way of saying that any given time in history is the sum of all the events that preceded it, and therefore impossible to repeat in exactly the same way. The second assumption is, that anatomical features that can be traced through the fossil record to the older strata represent the primitive condition of the organisms. Thus, faunal succession combines the laws of history with the law of superposition to provide a reliable key to the stratigraphic position of the rock layers.

THE GEOLOGICAL TIME SCALE

The principles described above have been combined through application to develop the chronology of geologic events and organic succession as shown in Table 4-1. Many more facts are known than are detailed in the diagram, which is provincial to North America, and emphasizes familiar life forms. A simplified representation of a typical fossil is given for each period.

EXERCISES IN HISTORICAL GEOLOGY INTERPRETATION

The following exercises will draw upon your knowledge of elementary paleontology, rock origins, and the principles of stratigraphy. In geologic interpretation, try to keep in mind the basic meaning of the principle of *Uniformitarianism*; that the present is the result of the past,

74 Exploring Geology

and that geologic processes have remained essentially the same throughout geologic time. Continents and ocean basins may move about the face of the Earth, but limestones still form primarily in two or three environments, and reverse faults are always the result of compressional forces. Drawings of representative fossils are included here for comparison with laboratory specimens and for working out the stratigraphic range exercise.

[handwritten note: ⊥⊗ = strike/Dip related to original horizonality. need direction (cardinal pts.)]

Fossils 77

NORTH AMERICAN TECTONICS	REPRESENTATIVE UNITS
Western U.S. shears against Pacific Plate causing widespread tectonism. Appalachians rejunvenated. Equator achieves present day position.	Scattered volcanism (Crater Lake formed, Mt. St. Helens erupts repeatedly); alluvial deposits in valleys. Mississippi delta forms, Atlantic coastal sediments deposited.
Batholithic intrusion and composite volcanic activity continue to accompany subduction in the west. The continent is completely emergent, with coastlines much like today. Modern physiographic provinces develop. Atlantic completely open.	Wasatch Formation (pink sandstone cliffs of Bryce Canyon). Volcanics in Yellowstone, Columbia Plateau, Cascades, San Juan Mts., CO. Isolated sialic plutons, Green River, WY. Lake beds (oil shales) famous for fossil vertebrates. Goldbearing veins of the Black Hills. Continental deposits of lignite and bituminous coal. Oil and gas in Gulf coast, southern California and offshore. Diverse metallic ores of Colorado Rockies.
Early in this period, the continent is mostly emergent, but later the western seaway floods into the mid-continent leaving an emergent mountain chain formed by the force of Pacific Plate/North American Plate collision. The Atlantic Ocean is open except at northern end.	Formation in Rocky Mtns. and to the east include primarily marine Dakota Sandstone, Mancos Shale, Mesaverde group, Fox Hill Sandstone; The Dakota is an aquifer, and some of the sandstones contain coal. Intrusion of Sierra Nevada and other sialic plutons. Gold-bearing veins of Sierra Nevada. Possible barrier reef off Atlantic coast. Extensive limestone formation along Gulf coast.
At first, the continent is mostly emergent, with volcanic islands in the western seaway. Later, the seaway expands into the central continent. Granitic intrusions from subduction in the west.	Navajo cross-bedded sandstone of the Colorado Plateau. Deep-water marine Franciscan group of northern California. Morrison Formation of Colorado Plateau and Rockies. Gulf coast acquires its present form; salt deposits in Gulf of Mexico. Mafic intrusive (now serpentinite) and andesites in California.
Eastern and central North America are above sea level. In the far west, shallow sea bottom surrounds island volcanoes, and fault-controlled basins and ranges develop offshore. The southern Atlantic Ocean begins to open up. (See Figure 4.1.)	Felsic to mafic intrusives from Oregon to Alaska. Newark red beds, Palisade basalts, Moenkopi/Chinle red sandstones and shales of the Colorado Plateau. Chinle in Arizona contains agatized tree trunks of the petrified Forest. Dockum red beds in Texas.
Eastern U.S. largely emergent land. Volcanic islands indicate continuing subduction in the west. Mid-continent regions collect carbonates, evaporites, and terrestrial red sandstones and shales. Pangaea separates; North America begins to move westward.	Upper 2,000 feet of the Grand Canyon including Kaibab Limestone rimrock. Phosphoria of northwest U.S. Dunkard red beds in West Virginia. Extensive reefs in Texas and New Mexico, including limestone of Carlsbad Cavern. Extensive salt deposits in Kansas.
Northern proto-Atlantic Ocean closes, completing uplift of the Appalachians and Ouachita, formation of the supercontinent Pangaea. Volcanic activity along western continental margin. Extensive coal deposits formed.	Fountain Arkose of Rocky Mountain Front Range; Paradox Basin evaporites of Utah. Metamorphic Calaveras sequence of western Nevada to Alaska. Cyclic deposition of coal beds from northern Illinois to West Virginia.
Proto-Atlantic ocean closes in the southern part as Africa collides with southeastern North America, creating the southern Appalachian Mtns. Widespread shallow seas collect limestone and shale.	Many cliff-forming limestones in the Rocky Mountains, Mississippi Valley, and southern Appalachians. Clastic sedimentary units followed the carbonates.
Emergent mid-continent areas are drowned later in the period by shallow seas. North America and Europe again collide, pushing up Appalachians. Continuing subduction in far west creates north-south mountains from Nevada through Canada.	Temple Butte limestone of Grand Canyon. Devil's Gate carbonates, Nevada. Oil-bearing carbonate reefs of the Williston Basin in the northern Rocky Mtns. Extensive delta sediments of the Catskill Mountains.
Appalachian highlands and western U.S. volcanos are islands in North American shallow sea. Organic reefs ring mid-continent basins collecting evaporites (gypsum, halite).	Limestone of the Silurian Hills, Mojave Desert; some fossiliferous units in the northern Sierra Nevada and Klamath mountains. Granitic intrusions in northern Appalachians, Niagara Falls series of limestone and shale. Michigan Basin evaporites.
Gentle warping of Precambrian basement and sediment cover to form basins and arches. Shallow seas still cover most of North America. Ancestral Pacific Ocean floor is being subducted under western North America. North America and Europe push together uplifting Appalachians.	Granite and marine basalts in Alaska produced by subduction. Pure quartz sandstones moved westward from emerging crystalline basement to form Eureka Quartzite of Basin and Range. Harding sandstone of Colorado, deep water melange sediments in Vermont and Newfoundland. Deep water marine shales in far west and east.
Partially emergent Precambrian crystalline rocks centered on Hudson Bay region are surrounded by marine waters grading from shallow to deep near present coastlines. Near-shore sandstones are flanked by shallow limestones and deeper water shales. Equator bisects North America from north to south.	Marine sandstones, as the Tapeats of Grand Canyon, the Prospect Mtn. Quartzite of Basin and Range, and the Potsdam of New York. Marine carbonates, as the Noonday Dolomite of southeast California. Marine shales along the western and eastern margins of the continent.
Continental crust is formed by mantle differentiation. Two glaciation events. Southern continents clumped together (Gondwanaland); Europe and North America close together.	Stillwater, MT chromite deposit. Grand Canyon inner gorge crystalline and sedimentary rocks. Great iron ore deposits on Lake Superior. Sudbury nickel-copper deposits.

Table 4-1b. The Geologic Time Scale

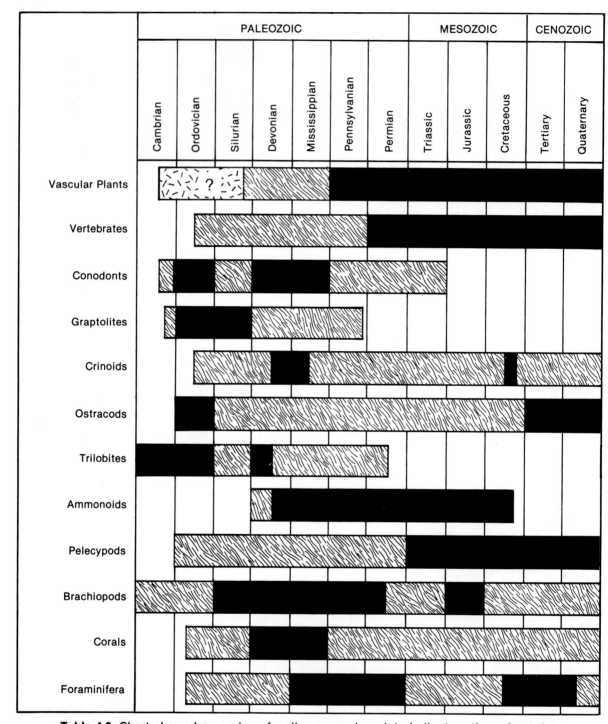

Table 4-2. Chart shows how various fossils are used as date indicators throughout the geologic time scale. The black areas show the time of maximum utility for correlation, the wavy lines indicate the total geologic range.

PHYLUM PROTOZOA

Foraminifera

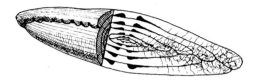
Fusulinella
Penn.

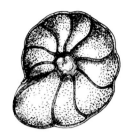
Cibicides Pachyderma
Eoc.

Textularia
Jur.-Rec.

PHYLUM COELENTERATA

Solitary Corals

Zaphrenthis
Dev.

Amplexizaphrentis, Sp.
Miss.-Penn.

Colonial Corals

Halysites
Ord.-Sil.

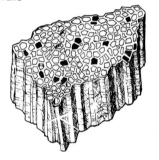
Favosites
U. Ord.-M. Dev.

PHYLUM BRACHIOPODA

Mucrospirifer
Dev.

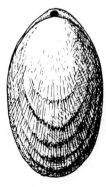
Terebratula
Cret.-Tert.

Echinoconchus
Miss.

Rhynchotreta
Miss.

Figure 4-1. Fossils

80 Exploring Geology

PHYLUM MOLLUSCA

Pelecypods

Venus berryi
Mioc.

Exogyra
Jur. Cret.

Trigonia
Jur. Rec.

Gastropods

Euomphalus
Miss. M.-Trias.

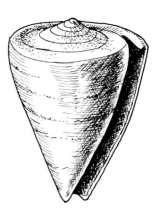
Conus
Eoc.-Rec.

Cephalopods (Ammonites)

Meekocera gracilitatis
Trias.

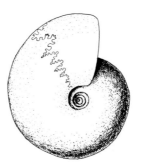
Ceratites
M. Trias.

Lytoceras
Jur.-Cret.

Figure 4-2. Fossils

Fossils 81

PHYLUM ARTHROPODA

Trilobites

Ostracods

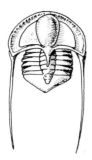

Nevadia weeksi
L. Camb.

Cryptolithus
Ord.

Phacops
Sil.-Dev.

Tetradella
Ord.-Sil.

Primitia
Ord.-Perm.

PHYLUM ECHINODERMATA

Echinoid

Crinoids

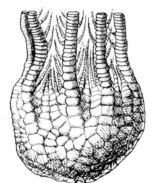
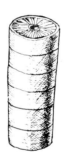
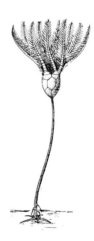

Clypeaster grandis
Rec.

Uintacrinus
U. Cret.

Crinoid stem
fragment

Agassizocrinus dactyliformis
U. Miss.

PHYLUM CONODONTA

Loxodus
L. Ord.

Drepanodus
Ord.-Sil.

Polygnathus
Dev.-L. Miss.

PHYLUM HEMICHORDATA (Graptolites)

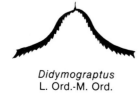

Didymograptus
L. Ord.-M. Ord.

Monograptus
Sil.

Figure 4-3. Fossils

Fossils 93

Name _____

Section _____

FOSSIL IDENTIFICATION QUIZ #1

	Quaternary									
	Tertiary									
	Cretaceous									
	Jurassic									
	Triassic									
	Permian									
	Pennsylvanian									
	Mississippian									
	Devonian									
	Silurian									
	Ordovician									
	Cambrian									
Stratigraphic Range Useful for Correlation	Fossil Identification									

Name _____

Section _____

FOSSIL IDENTIFICATION QUIZ #2

	Quaternary									
	Tertiary									
	Cretaceous									
	Jurassic									
	Triassic									
	Permian									
	Pennsylvanian									
	Mississippian									
	Devonian									
	Silurian									
	Ordovician									
	Cambrian									
Stratigraphic Range Useful for Correlation	Fossil Identification									

96 General Geology

Physiographic Regions and Provinces of the Coterminous United States

(U.S.G.S. Photo)

#1: Pacific Border Province; #2: Cascade Sierra Nevada; #3: Basin and Range Province; #4: Colorado Plateau; #5: Columbia Plateau; #6: Northern and Central Rocky Mountains; #7: Wyoming Basin, #8: Southern Rocky Mountains; #9: The Great Plains; #10: Interior Lowlands; #11: The Canadian Shield; #12: Appalachian Trend; #13: Atlantic/Gulf Coastal Plains; #14: Continental Shelf.

5

Topographic Maps

INTRODUCTION

A well-constructed topographic map is an excellent way to present many different kinds of information. The informed reader can accurately determine the location of a point on the Earth, the height of a mountain, the distance between two points, the surface area of a lake or city, the steepness of a hiking trail, the distribution of water and vegetation, road placement, and the shape and character of the land itself.

The purpose of this chapter is to acquaint you with the theory and language of topographic maps necessary to *prepare and read* them. Developing map-reading skills will make you more comfortable with using maps for driving, wilderness exploration, locating property accurately, and many other activities. For those planning careers in the areas of earth science, marine science, engineering, land development, city planning, water resources, environmental studies, aviation and the miliary, knowledge and understanding of topographic maps is essential.

The principles discussed here will be applied in later exercises on *geologic maps*, and *applied geology*.

MAP TYPES

You may have had occasion to sketch directions to your house for a friend, showing streets, local landmarks, and the desired destination. Such a map is a **planimetric** map, as are road maps published by oil companies and automobile clubs, political maps, and even so-called treasure maps. These maps portray the location of cultural and natural features such as roads, cities, political borders, rivers, shorelines, and mountain peaks in a two-dimensional display. Planimetric maps vary in detail and accuracy, and are generally designed for destination-oriented tasks or to delineate land-control boundaries.

For more technical use, such as in the career areas outlined above, the portrayal of *relief* and *topography* (land elevation and shape) is essential, in addition to planimetric information. A **topographic map** contains *contour lines* with specific elevations which demonstrate the *vertical* (or third) *dimension* of the land, in addition to the planimetric data. Topographic maps therefore maximize available information, and it is with these maps that we will be concerned for most of this chapter.

MAP PROJECTIONS

Because the Earth is a sphere, all maps printed on a flat piece of paper are distorted in some respect. The map *projection*, or type of inherent distortion will be determined by what the map

98 Exploring Geology

is to be used for, and what area of the globe it represents. The **Mercator** projection, used chiefly in navigation, assumes that lines of latitude and longitude are always perpendicular; therefore, this type of map becomes progressively more distorted toward the poles, where land masses appear at several times their actual relative size. An **equal-area** projection maintains correct relative sizes of areas, but distorts their shape; a **conformal** projection has just the opposite effect. One of the best compromises is the **polyconic** projection used by the United States Geological Survey (U.S.G.S.) in making topographic maps.

MAP GRIDS

One of the primary functions of a map is to establish the location of a point or an area so that this position may be found at a later time, communicated in writing or speech, or recorded for legal purposes.

The most efficient way to establish location is by means of a *grid* or *network of reference lines* crossing at right angles which are numbered from some designated point. A person giving directions such as *"two blocks north and one block east from here,"* is using intersecting streets for a grid, and the location at which he is standing as a designated point. Similarly, the abscissa (x) and ordinate (y) axes of the Cartesian coordinate system make up a grid used for plotting (x, y) points, with the origin (0,0) as designated point.

Topographic maps primarily employ two grid systems: **latitude** and **longitude**, which is an international location net using the equator and Prime Meridian as reference lines; and **township** and **range**, used in the United States to designate land ownership and control. Each system is further explained below.

Lines of **latitude** (or parallels) are rings around the Earth parallel to the equator, which is 0° latitude. Because there is one quarter of a circle (90°) from the equator to a pole, lines of latitude are numbered from 0° to 90° north or south of the equator. Lines of **longitude**, also called meri-

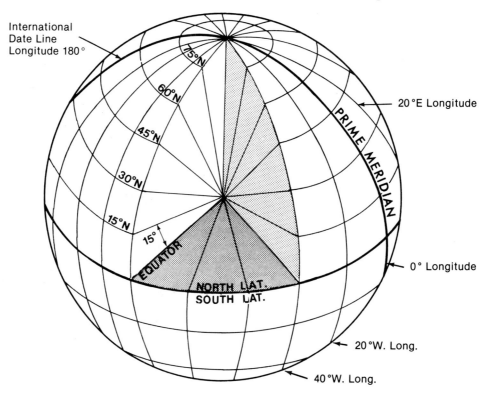

Figure 5-1. Latitude and Longitude Grid System

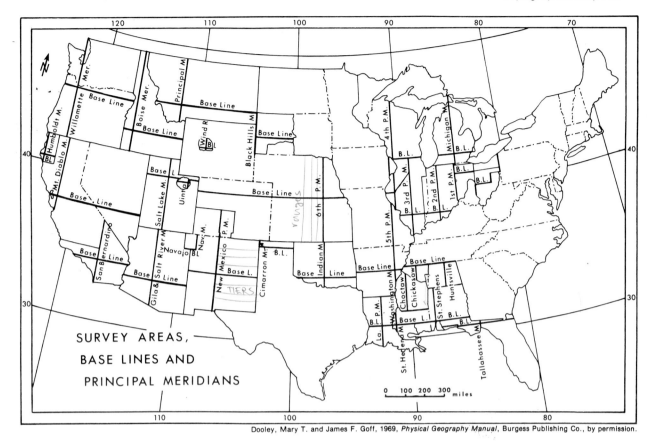

Figure 5-2. Principal Base Line and Meridian Systems

dians, are circles which pass through both poles. The Prime Meridian, 0° longitude, runs through Greenwich, England, and longitude is counted in degrees eastward and westward to 180° at the International Dateline. Latitude and longitude thus form a grid on the Earth's surface which can be used for locations on land or water. See Figure 5-1 for a graphic presentation of this grid. All professionally produced maps are oriented with north to the top.

Latitude and longitude lines are circles, which is why they are calibrated in *degrees*. A degree can be broken down into smaller units of *minutes* and *seconds* in the same manner that the time unit of one hour is subdivided. There are 60 minutes (') in a degree (°), and 60 seconds (") in a minute. The correct designation for 1½ degrees would be 1° 30'; adding one-quarter minute to this would give 1° 30' 15".

In order to rectify vague and provincial methods of designating property boundaries, and to bring uniform order to legal land descriptions, the General Land Office in 1785 established a grid system commonly referred to as **Township** and **Range**. Everyone who owns property or deals with property descriptions should be aware of this method of land location.

As the United States was systematically surveyed, township and range guides were established; they were created by the intersection of the **Base Line** corresponding to a line of latitude, and a **Principal Meridian** or line of longitude (Figs. 5-2, 5-3). Six-mile-wide strips of land parallel to the Base Line, called **tiers**, are numbered consecutively north and south. Six-mile-wide strips of land paralled to the Principal Meridian, called **ranges**, are numbered consecutively east and west.

Each square resulting from the intersection of a tier and range is called a **township**. It is six miles on each side (36 square miles) and is designated by its tier number north or south, and its range number east or west. For example, the shaded township in Figure 5-3 is T2S, R4E. Every

24 miles north and south of the Base Line a correction line of offsets is established to take up the convergence of the meridians towards the poles (see Fig. 5-3, line xy). These offsets can cause errors to be made in establishing property lines.

Each township of 6 by 6 miles is divided into 36 **sections** (Fig. 5-4) of one square mile each and these are numbered starting in the upper right corner, going across the top to the left and then back and forth in snake fashion to the lower right corner. The purpose of this numbering system is to assure that consecutively numbered sections will be adjacent.

If you were to buy a quarter-acre lot, you could hardly designate it as a township or even as a section, since one section is 640 acres. So a system of quartering was devised whereby it is possible to break a section of 640 acres down into quarters of 160 acres each, and those into quarters of 40 acres each, and those into quarters of 10 acres each, and finally the 10-acre parcels can be quartered once again into 2½-acre parcels (Fig. 5-5). Generally, a parcel of property smaller than 2½ acres is further delineated as a half of the quarter.

The usual order in which property designations using this system are given is the reverse of the way it is figured out, starting with the smallest parcel. The shaded 10-acre parcel in Figure 5-5 is designated as:

> SE¼, NE¼, SE¼, Section 19, T2S, R4E of (for instance) the San Bernardino Base Line and Meridian.

It should also be noted that some parts of the United States do not employ the township and range grid for property location. Texas and some other areas which were originally Spanish land grant tracts do not use the township and range system. The area of the original thirteen colonies uses the system of metes and bounds.

Topographic Maps 101

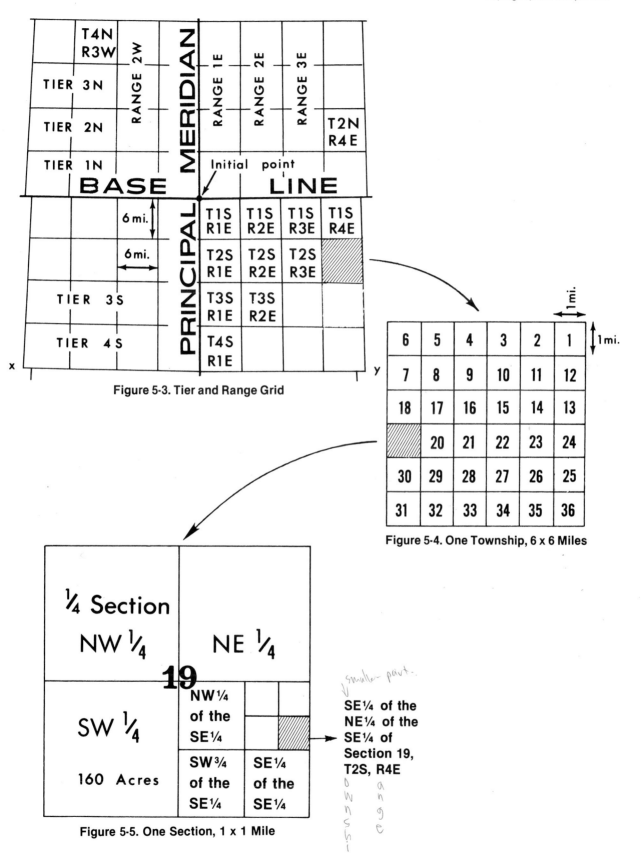

Figure 5-3. Tier and Range Grid

Figure 5-4. One Township, 6 x 6 Miles

Figure 5-5. One Section, 1 x 1 Mile

SE¼ of the NE¼ of the SE¼ of Section 19, T2S, R4E

102 Exploring Geology

MAP SCALES

The fundamental usefulness of maps derives from the fact that they represent the reduction of vast areas down to a piece of paper we can easily handle. In order to interpret a map successfully, we must know the amount of this reduction, and know how a unit measured on the map relates to actual distance on the ground.

The amount of reduction is expressed on maps as the **ratio scale**; i.e., the distance on the map is a certain fraction of that on the ground. Note that any system of units may be used, as the ratio scale is unitless. A 1:24,000 scale means that any one linear unit measured on the map is equal to 24,000 of those units on the ground. Using one inch for the unit, one inch on such a map would represent a distance of 24,000 inches, or 2,000 feet, on the ground. So for all maps printed at a ratio scale of 1:24,000, the relation of map to ground units, the **verbal scale**, is one inch equals 2,000 feet, 1″ = 2,000′.

A graphic scale or **bar scale** is a plot of the verbal scale on the map. This scale, which is printed at the bottom center of all U.S.G.S. maps, is actually a ruler for measuring map distances. This scale will still be valid if the map on which it is printed is reduced or enlarged photographically, although the ratio scale will not. A 1:62,500 bar scale looks like this:

CONTOUR LINES

Contour lines are used to depict three-dimensional features on a flat piece of paper. Contours show the shape of hills, mountains, and valleys, as well as their altitude. A **contour** is an imaginary line on the ground, all points of which are at the same altitude or, put another way, *a contour line is a line connecting all points of equal elevation*. The zero contour is the shoreline of the ocean halfway between high tide and low tide (mean sea level). All points 10 feet above sea level would lie on the 10-foot contour line; all points 20 feet above sea level would lie on the 20-foot contour line, and so on. In this example the *contour interval*, which is the difference in elevation between two adjacent contours, is 10 feet. A contour interval is chosen to fit the relief of the landscape and the scale of the map; to show as much relief as possible without cluttering the map with lines bunched too closely together. Commonly used intervals are 5, 10, 20, 25, 40, 50, 80, and 100 feet.

Figure 5-6 shows contour lines drawn on a natural landscape. If this imaginary area complete with contour lines were photographed from above, the resulting photo would be a topographic map. In fact, modern topographic maps are created by sophisticated computer processing of vertical aerial photographs.

Listed below are some rules summarizing the basic nature of contour lines which should be used when constructing or interpreting a topographic map:

1. On truly level ground there are no contours.
2. The spacing of contours reflects the gradient or slope:
   a. contour lines that are far apart indicate a gentle slope
   b. contours that are close together indicate a steep slope
   c. contours that merge indicate a vertical slope
3. Contours never cross, never branch, and never terminate.
4. All solid-line contours are multiples of the contour interval; e.g., if the contour interval is 10 feet, the contours will be 10, 20, 30, 40, 50, etc. Usually every fifth contour, called an *index contour*, is printed heavier for ease of reference.
5. Dashed contours represent elevations of half the normal interval, and are added in areas of low relief to increase detail.

6. Contour lines crossing stream valleys or other channels form a "V" pointing upsteam.
7. Jagged topography will make sharp angles in the contour lines; low, rolling landscapes will have gently curving lines.
8. Contour lines will eventually close on themselves, although demonstrating this may require consulting adjacent maps.
9. Normal contours enclose an area that is higher than the contour; i.e., all points that lie within such a closed contour are above the level of that contour.
10. Depression contours enclose areas that have no outlet— closed basins. They are marked by hachures on the inside:
11. Depression contours have the same elevation as the adjacent normal contour which encloses the depression.
12. Contours must be counted consecutively, and none can be skipped; repeated contours adjacent to one another indicate a change in slope direction, such as into a depression or across a stream.
13. Bench marks, points on contours, and spot elevations are exact.
14. Hilltop elevations may be estimated as being greater than the highest contour shown but less than the next contour (imaginary) above. Similarly, the bottoms of drainages will be below the lowest contour shown in the immediate area, as will the bottoms of depressions.
15. The elevation points between contours may be estimated by the position of the point. If the point is midway between two contours, the elevation will be read as halfway between the values of the two bracketing lines. The nearer the point is to a contour, the closer it is to the elevation of that contour.

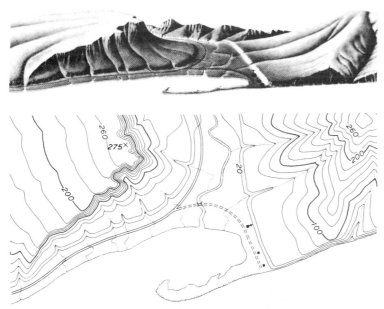

Figure 5-6. Landscape with Contours

U.S. GEOLOGICAL SURVEY TOPOGRAPHIC QUADRANGLES

Standard U.S.G.S. quadrangles are maps of small sections of the Earth and are bounded by the same amounts of degrees or minutes for both latitude and longitude. They approach a true square only at the equator, since lines of longitude converge toward the poles. The sizes of maps printed by the Survey, their relation to each other, and the scales for each are given in Figure 5-7. You will primarily be working with 7½' and 15' quadrangles. Note that a 15' quadrangle will depict four times the area of a 7½' quadrangle, and yet may be a smaller map.

104 Exploring Geology

A number of important details are given in the margins of these maps. Compare Figure 5-8 with one of the 15' quadrangles found in the back of this manual in order to familiarize yourself with this information described below (clockwise from top right):

1. *Top right:* Name, location and size of the quadrangle.
2. *Bottom right:* Name, date of publication, road classification.
3. *Toward bottom center:* Quadrangle location on state map.
4. *Bottom center:* Scales applicable to quadrangle, contour interval.
5. *Left of center:* Magnetic declination. The Earth's magnetic and rotational poles do not exactly coincide, so magnetic compass readings in most places will deviate from true north. Magnetic readings must be corrected by the amount given to the right of the symbol, and in a direction opposite to the declination.
6. *Bottom left:* Details of how the area was mapped and who did the surveying, if mapped before the 1940's when aerial photography became the principal method of mapping.
7. At all corners and center of each side, in parentheses, is the name of the adjoining quadrangle and its scale if different from the map you are reading.
8. Around the periphery you will find township and range data printed in red, and intermediate minutes of longitude and latitude between corners in black.

Standard symbols and colors used on these maps will be found on the page opposite p. 140; take some time to look them over. Most of the symbols attempt to simulate what they represent (churches have crosses, etc.). Water is shown in blue, ground relief in brown, roads and buildings in black, urban areas in pink and purple, and vegetation in green.

DATA FROM TOPOGRAPHIC MAPS

Many kinds of information may be derived from topographic maps by using simple measurements and arithmetic.

Measuring Distances: If the distance to be measured is a straight line, simply mark it off on the edge of a piece of paper and compare it to the bar scale. If the distance being measured crosses onto a map of a different scale, figure each separately and add the results to get full distance.

For irregular or curved distances, such as that along a stream course, place the corner of a sheet of paper at the beginning point and line the edge up against the first leg of the route. Put a pencil point at the end of the leg and swing the paper edge to conform with the next segment. Continue this, averaging out minor irregularities on the route, until you reach the other end of the distance being measured. Mark the end, and then compare the length of the edge from corner to end with the bar scale.

Relief: Relief is the difference in elevation between the lowest and highest points within the area. Look for the high point along divides and at mountain peaks, and the low point along a stream or body of water. Remember to check spot elevations and unmarked peaks.

Computing Gradient: Gradient is a measure of the slope of the land. It is expressed as a ratio of total vertical drop between two points per unit of horizontal distance. For example, a certain stream drops through three 10-foot intervals along a distance of 2½ miles. To calculate gradient, solve for X in the following equation:

$$\frac{30 \text{ ft}}{2.5 \text{ mi}} = \frac{X \text{ ft}}{\text{mi}}$$

X = 12. The gradient is 12 ft. per mile.

Topographic Maps 105

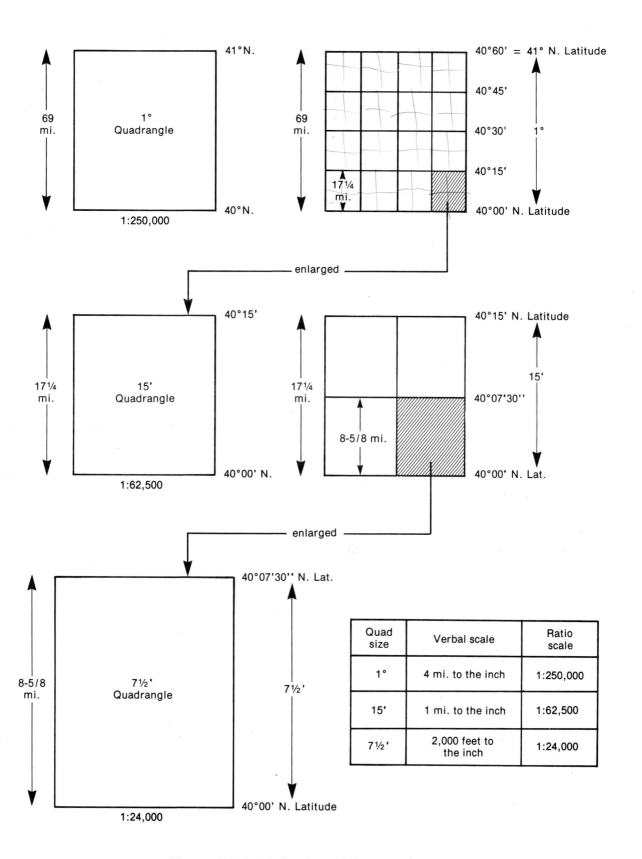

Figure 5-7. U.S.G.S. Quadrangle Sizes and Scales

106 Exploring Geology

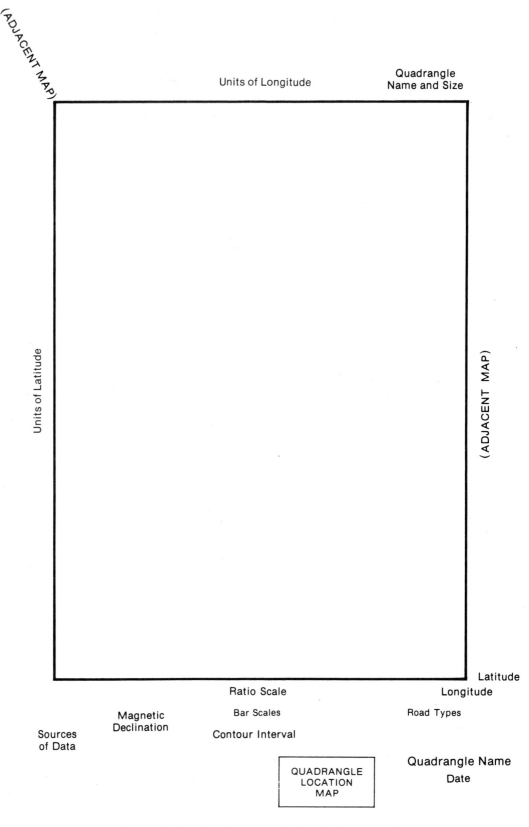

Figure 5-8. U.S.G.S. Quadrangle Marginal Information

Making a Contour Map: Making a complete contour map from seemingly sketchy initial data is quite a challenge, but it is rewarding to see the shape of the land develop from isolated numbers. You will start with a base map showing spot elevations, drainages, and water bodies. Interpolating between given elevations, find as many points as possible which correspond to full contour intervals. Then begin to connect points of the same elevation. It is best to start in one area of the map bounded by drainages and establish all contour lines within that area. Carry them forward to the next area as a group. Remember upstream "V"s, the rules for depressions, and other contour rules as outlined on pages 102-3. In this type of exercise, where there is no opportunity to field-check the contours, variations in data interpretations are expected, and each student's effort will be somewhat different. Any interpretation is acceptable which does not break the rules of contours.

Making a Topographic Profile: A **profile** is a silhouette of part of the Earth's surface; it is an irregular line showing the surface in cross-section. In the exercise on making a topographic map and profile (page 133), complete the topographic map and then draw a line from A to A'. Wherever that line crosses a contour, drop a perpendicular down to the profile line of corresponding elevation and make a dot. You should also drop a perpendicular from each hilltop and stream bottom. When all elevations have been projected downward, connect the dots to complete the profile. Be sure that the profile extends all the way to the end of the A-A' length. Remember that streams run in gullies, that hilltops are usually rounded, and attempt to make your profile as realistic as possible.

In making a profile from a standard topographic map place the top edge of a piece of graph paper along the line where the profile is to be made. Mark all contours, streams, and hilltops as small tick marks along the paper's edge and label their respective elevations. Then set up the profile lines on graph paper and drop perpendiculars down vertically to the appropriate elevation levels previously established. Connect the points as above.

Vertical Exaggeration of Profiles: Horizontal map distances are usually a different order of magnitude than the vertical relief along the same distance. Ten miles of horizontal distance may show only a few tens of feet of vertical change, and a profile to the scale of the map would most closely resemble a straight line. For this reason, topographic profiles are usually stretched vertically, or **exaggerated**, in order to emphasize relief. This distortion is a valid technique as long as the amount of stretching is stated. After drawing a profile, you should be able to calculate the vertical exaggeration, as in the following example:

A profile on a 7½' quadrangle (1:24,000, or 1"=2,000') is plotted with a vertical scale of 1"=500'. The calculation of the vertical exaggeration is:

$$\text{Vertical Exaggeration} = \frac{\text{Vertical scale}}{\text{Horizontal scale}} = \frac{1''/500'}{1''/2,000'} = \frac{2,000}{500} = 4X$$

Exercises #1 and #2
Cayucos, CA Quadrangle

CAYUCOS, CALIFORNIA 15'
U.S.G.S. TOPOGRAPHIC MAP

Cayucos, CA
This easterly view of the Morro Bay area delineates a number of geologic features. Morro Rock in the left foreground can be seen as an eroded portion of a plutonic trend which continues to the upper right of the photo. The Morro Bay State Park sand bar is prominent in the foreground. (Photo by John Shelton)

The Cayucos, California 15-minute quadrangle has been included as a complete quadrangle and two sets of exercises on this map follow. The first set is concerned with the different kinds of information included on U.S.G.S. quadrangles, and the second with reading and interpreting contour lines.

Morro Rock, Black Hill and other peaks to the southeast consist primarily of Oligocene dacite porphyry intruded as a sheet or series of plugs along a tectonic lineament. The plutonic rocks have been etched into prominence by selective erosion of surrounding older, softer materials. These older rocks are part of the Franciscan Formation, a massive unit of deep water marine sandstone, shale, and mafic volcanic rocks which continues north and south of the bay. Franciscan rocks were thrust up onto the continent as a result of the interaction of the Pacific ocean floor and the westerly moving North American continent. They are locally sheared and metamorphosed as a result of this tectonic activity.

Faults in the area roughly parallel the coastline and have elevated some of the weaker Franciscan rocks which has led to prehistoric and modern landslides and mudflows. Exercise questions are on pages 119 and 121.

110 Exploring Geology

Exercises #3 and #4

Refuge, ARK-MISS, Quadrangle

REFUGE, ARK-MISS 15'
U.S.G.S. TOPOGRAPHIC MAP

Greenville, Mississippi
Note how different tones of grey show different soil types. Match the sand bars, filled in lakes, natural levees, and river channels with their topographic designations.
(Photo U.S.G.S.)

    The Mississippi River dominates the central United States and is the major topographic and geologic feature of the Refuge, ARK-MISS quadrangle. Virtually every feature on the map is related to either the changing course of the river or to the sediments deposited by it. Largest of the rivers in the United States, the Mississippi drains a third of the nation, discharging 630,000 cubic feet of fresh water per second into the Gulf of Mexico. It also carries millions of tons of sediment, much of it topsoil from the farmlands of the Midwest, into the Gulf annually. Historically, the great river has brought devastation from its floods, and incredible fertility to its valley by the rich soil left behind after the floods. It provides an inland waterway into the nation's food producing areas, but requires constant work to keep the channels open. It is a remarkable example of the dynamic systems that operate on the Earth.

    At the end of the Cretaceous Period the great shallow seas that had covered most of the mid-continent retreated, leaving the land emergent and subject to erosion. At that time the Mississippi reached the sea just south of Cairo, Illinois. During the Tertiary Period thousands of cubic yards of sediment were deposited southwards, filling in the Gulf of Mexico and building up the land surface. Beneath the Mississippi River and most of the Gulf Coast lie upwards of 20,000 feet of sand, silt, and clay brought by the rivers that drained the mid-continent after the Cretaceous Period.

    During the Pleistocene glacial epochs the Mississippi was choked with sediments and, because sea level was about 200 feet lower than it is now, deep canyons were cut into its valley. With the end of the ice ages and a corresponding rise in sea level, the river channels were filled with sediment and the Mississippi became a meandering stream. Meandering rivers have shallow gradients and broad valleys. They erode their channels on the outside (convex) margin of curves and deposit sediment on the inside (concave) banks of the curves. This system provides an effective way of transporting sediment downstream and also enlarges the river valley by eroding the banks. Sediment that is deposited on the inside of the curve builds up sand bars with coarser materials at the bottom (the bed of the river) and finer particles near the top. These build out into the stream bed as the opposite bank is eroded and the curve, or meander, is extended laterally.

During periods of flooding the river overflows its banks, rapidly losing the velocity that enabled it to keep sediments in suspension. The particles that drop out build up to form a natural levee along the banks of the river. Eventually curves become so tight that the river breaks through the neck of the meander and a new cycle of meander formation begins. The old curve becomes an oxbow lake. It is replenished with water during flood stages and is gradually filled in with fine-grained sediments and vegetation.

The river changes constantly. The channel swings from side to side, sand bars form and disappear, meanders migrate, enlarge and are cut off. For those who use the river and its valley for farming, navigation, or to establish borders, a constant effort is needed to keep pace with the dynamics of a living river system.

Note the changes in the state boundary between Arkansas and Mississippi. Find the sand bars and order them chronologically. What kind of soil would you expect to find in oxbow lakes? How wide is the fertile valley of the river? What special problems face the various government administrators? Exercise questions are on pages 123 and 125.

112 Exploring Geology

Exercise #5
Ennis, MT Quadrangle

ENNIS, MONTANA 15'
U.S.G.S. TOPOGRAPHIC MAP

Ennis, Montana

This is a vertical aerial photograph of the fault-controlled Madison Range mountain front near Ennis, Montana, and the broad low alluvial fan extending to the braided Madison River.
(U.S.G.S. Photo)

This map shows a portion of the Madison Range and adjacent river valley in the Rocky Mountains of western Montana.

The topography is controlled by normal faulting of mid-Tertiary age. Prior to this uplift, the region was eroded almost to sea level, and the Madison River flowed south into the Snake River drainage in Idaho. After uplift, the river was forced back to the north where it initially ponded up to form Ennis Lake. The lake lowered considerably as the river cut a gorge to the north draining off much of the water. Evidence of Pleistocene glaciation may be seen in the mountains; alluvial deposits eroded from them are currently filling the valley. Exercise questions are on page 127.

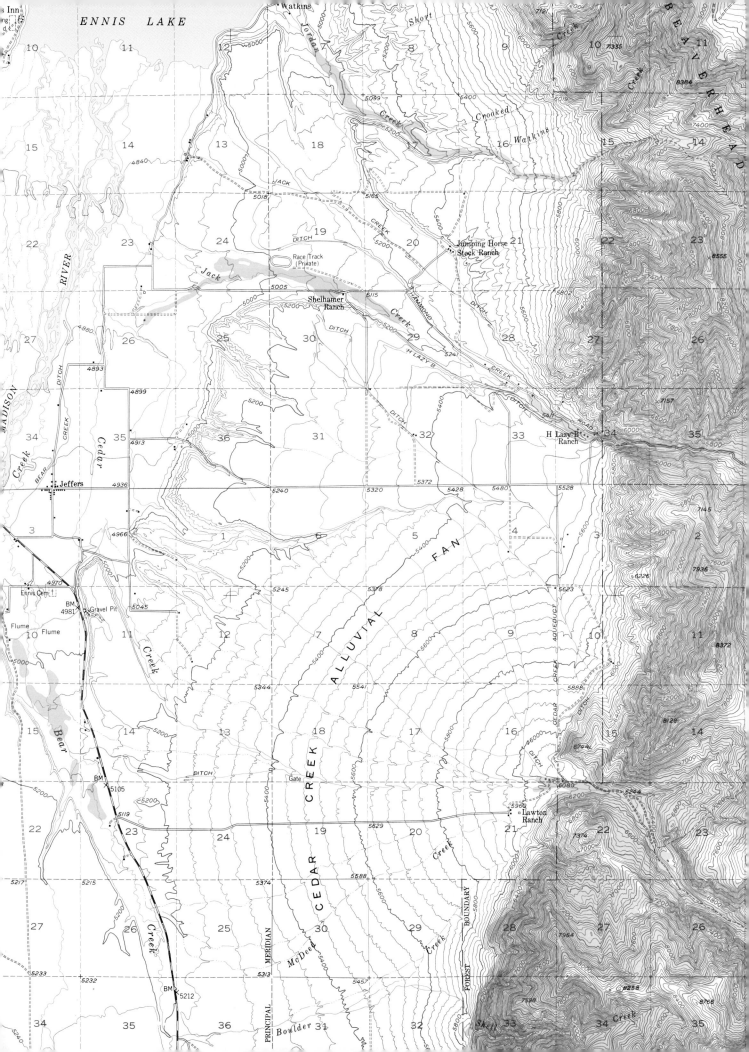

114 Exploring Geology

Exercise #6

Cordova, AK Quadrangle

CORDOVA D-3, ALASKA 1:63,360
U.S.G.S. TOPOGRAPHIC MAP

QUADRANGLE LOCATION

Cordova D-3, Alaska

This photo shows alpine (valley) glaciers in Alaska with prominent moraines. The stripes of rubble picked up by glacial abrasion and avalanching extend downstream parallel to the direction of flow.

(U.S.G.S. Photo.)

Glaciers on Earth at the present time are not as extensive as they were 10,000 years ago. Many good examples of glaciated areas may still be found at high altitudes and high latitudes, however. On this sheet can be seen a number of alpine (or valley) glaciers, those ice sheets which are constrained to flow within former stream channels. Blue lines represent contours on the surface of the ice. The brown speckled pattern delineates visible rock debris which is being transported and deposited by the glaciers. Exercise questions are on page 129.

116 Exploring Geology

Exercise #7
Lavic, CA Quadrangle

LAVIC, CALIFORNIA 15'
U.S.G.S. TOPOGRAPHIC MAP

QUADRANGLE LOCATION

Lavic, California

This vertical aerial photograph of the Lavic area shows a number of typical Mojave Desert features, including Holocene basalt flows, a dry lake, and a faulted alluvial fan.

(U.S.G.S. Photo)

    This quadrangle demonstrates many of the geologic features typical of the Mojave Desert, a wedge-shaped sub-province of the Basin and Range. The Mojave is bounded on the north by the Garlock Fault, and on the west by the San Andreas Fault. These two faults cross in the region of the Transverse Ranges. Because the Garlock Fault is left-lateral, and the San Andreas is right-lateral, the Mojave region has been moving to the southeast relative to surrounding areas. Rocks of many ages and origins occur here. Faulting and volcanism have prevailed from Oligocene to Holocene time.

    Topographically, fault-controlled discontinuous mountain ranges are aproned with several generations of alluvial fans which slope to areas with interior drainages leading to **playas** (intermittently dry lakes). Holocene lavas are shown here with a "wet sponge" pattern. Exercise questions are on page 131.

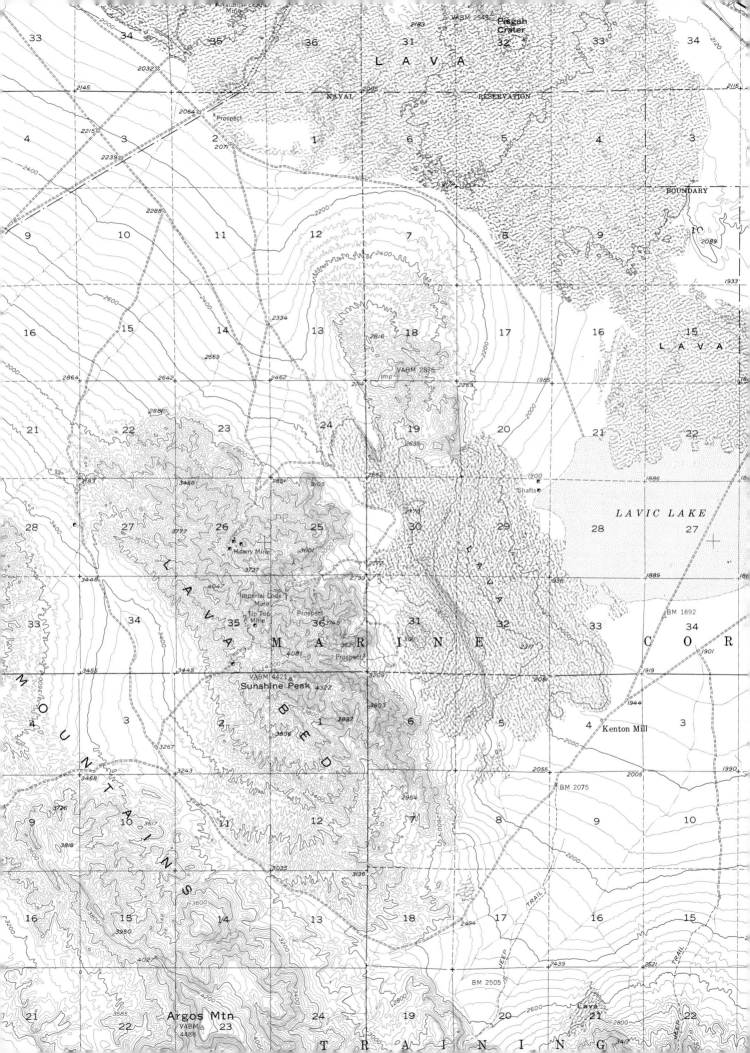

Name _____

Topographic Maps Exercise #5   Section _____
Ennis, Montana 15′ Quadrangle

QUESTIONS

1. What was the approximate elevation of the bottom of the larger lake which formerly filled the valley?

2. Estimate the minimum amount of vertical uplift which has taken place in this region since mid-Tertiary time.

3. Give the geological process responsible for creating each of these geomorphic features.

   Process

   a. Cedar Creek alluvial fan   _____

   b. Cedar Creek canyon   _____

   c. The tributary valley to Jack Creek in Sections 26 and 35
      along the east edge of the map   _____

4. What kind of rock material would you expect to make up Cedar Creek alluvial fan?

   Rock Family   *Rock Type*

   _____   _____

5. a. Compute the gradient of the Madison River from the edge of the map to Ennis Lake. _____

   b. Compute the gradient for Cedar Creek from the east side of the map to Section 18. _____

   c. What differences in configuration do you notice between low- and high-gradient drainages?

6. Several drainages on Cedar Creek alluvial fan originate well below the head of the fan (see Sections 7, 9, 20). What is the source of water for these drainages?

Name _____

Topographic Maps Exercise #6: Section _____
Cordova D-3, Alaska 1:63,360 Quadrangle

QUESTIONS

1. Glaciers characteristically widen and deepen stream valleys into broad, open ones that are U-shaped. Locate by general map area and section numbers an example of one of these valleys.

2. Why do the glaciers in the northern half of the map decrease in size to the east?

3. **Moraines** are piles of rock debris deposited by melting ice. They are generally named for their relative position in the glacier.
    a. Where are the best-developed terminal (end) moraines on the map?

    b. Does the medial (middle) moraine on Heney Glacier in Section 11 form a ridge or a valley?

    c. What specific processes might contribute rock debris to the lateral (side) moraines on Heney Glacier?

    d. What is the origin of the medial moraine on Heney Glacier?

4. **A cirque** is a formerly ice-filled bowl-shaped depression at the upper end of a U-shaped valley. Cirques are often occupied by small lakes called *tarns*. Locate, by section number, a cirque.

5. **Horns** are splendid isolated peaks which have cirques carved into them on several sides. Find two horns on the map, and locate them by section number and peak altitude.

6. There are several indications that glaciers were once more extensive in this area than they are at present. List and briefly discuss some of these.

Name _____

Topographic Maps Exercise #7: Section _____
Lavic, California 15' Quadrangle

QUESTIONS

1. How many episodes of volcanism can you document here? Explain.

2. Take a careful look at the nature of the contour lines dominating the Sunshine Peak Range and those on Argos Mountain.
   a. Describe the differences between them.

   b. What might account for these differences?

3. In what section do you find the crater from which Holocene lava in the center of the map emanated?

   _____ (section number)

4. The following questions concern relative age and should be answered *older/younger*:
   a. Lava which emanated from Pisgah Crater is _____ than Lavic Lake.

   b. Central Holocene lavas are _____ than Lavic Lake.
   c. A well-defined fault scarp is shown in Sections 30, 31, and 32 in the center of the map. The fault

      is _____ than associated Holocene lavas.

5. Considering the relationships established in question 4, indicate the correct age sequence of the five features below with consecutive numbers (1 = oldest, 5 = youngest):

   Sunshine Peak lavas _____   Lavic Lake _____   Central fault _____

   Pisgah lavas _____   Central Holocene lavas _____

Name_____

Chapter 5: Topographic Maps   Section _____

REVIEW QUESTIONS

1. The ratio scale of a standard U.S.G.S. 7½-minute quadrangle is _____

2. The verbal scale of a 7½-minute quadrangle is _____

3. Why are there no units associated with a ratio scale? _____

4. Maps supplied by oil companies and auto clubs are of what type? _____

5. Where on a Mercator projection would you find the most distortion? _____

6. Why is a 15-minute quadrangle at 50 degrees north latitude longer (N-S) than it is wide (E-W)? _____

_____

_____

7. Any point in North America will have _____ latitude and

   _____ longitude.

8. A very small contour interval would be used to portray what general type of land surface? _____

_____

9. How many square miles are there in a township? _____ A section? _____

10. If a map is reduced or enlarged photographically, is the ratio scale on it still accurate? _____

    Is the bar scale still accurate? _____

11. A map printed at 1:12,000 would have a verbal scale of _____ (English system).

12. A map printed at 1:100,000 would have a verbal scale of _____ (metric system).

13. What can you say about the elevation of a point which is enclosed by a normal contour? _____

    _____ A depression contour? _____

14. How many 15-minute quadrangles are there in a 1-degree quadrangle? _____

Chapter 5: Topographic Maps

Name_____

Section _____

MORE CHALLENGING QUESTIONS

1. How can you determine the scale of any quadrangle adjoining the one you are studying? _____
_____

2. What is a contour line? _____

3. What is a contour interval? _____
_____

4. Can you think of a situation in nature where contour lines *would* cross? Describe.

5. Assume that you are consulting a topographic map which is very old. What types of features have probably changed the most since the map was printed?

6. What is magnetic declination? _____
_____

7. Define vertical exaggeration. _____
_____

8. What kind of distortion in land shape is introduced in profiles which are exaggerated vertically?

9. Why do contour lines make a "V" in the upstream direction where they cross drainages?

Structural Geology 6

INTRODUCTION

Many lines of evidence support the concept of a dynamic, mobile outer Earth. Earthquakes and volcanic eruptions attest to current crustal activity; rocks bent or broken in earlier times demonstrate that such movement has been occurring throughout most of the Earth's history.

Rocks, like man-made materials, have limits to their strength, and can be **deformed** by the application of a *stress*, or *force*, which exceeds that strength. *Deformation*, or *strain*, means that a rock or rock body has been bent, broken, or moved from its original position. The amount of deformation will depend upon a number of factors, such as the strength of the rock, the amount of stress applied, duration of the stress, temperature of the rock, depth of burial, and amount of water present. The kind of deformation, or structure, will be determined by the orientation of stresses applied. **Compression** (squeezing), **tension** (pulling apart), **shearing**, and **uplift** and **subsidence** all occur in the outer Earth, and produce characteristic deformation.

This chapter is concerned with the geometry and terminology of the most common patterns of rock deformation or structure. Because the application of stress is an *event*, deciphering the consecutive development of structures in a given area will give a time sequence of events. This provides a framework for the geological history of that area. An analysis of rock types and their probable origin will complete the historical picture. Deducing the geologic history of an outcrop or a region is a satisfying exercise in itself and is absolutely essential to the exploration, discovery, and development of petroleum and mineral deposits. Petroleum fields are almost exclusively associated with deformed sedimentary rocks. In Saudi Arabia, oil and gas are concentrated in a series of crumpled layers deformed when the Arabian Peninsula pulled away from Africa and pushed into Eurasia. A slight change in tilt between sedimentary layers served to trap oil and gas over thousands of square miles of east Texas. Petroleum deposits in southern California are frequently associated with the shearing plate boundary of that region. Solid mineral deposits cannot be efficiently exploited unless their structural setting is understood, since this strongly affects the origin, shape, and extent of such bodies.

This unit also includes explanations and exercises dealing with basic geometric concepts in the three-dimensional portrayal of structures, as well as with the more common types of specific structures. Additional exercises will focus on deciphering the history of a variety of geologic settings. Structural geology subjects will be developed according to the outline on the following page.

144 Exploring Geology

I. Forces Producing Structural Deformation — 3 forces — compression
                                        — tension
II. Strike and Dip   ⊕ — horiz. bed, no dip.   — shearing
III. Types of Structural Deformation
    A. Folds
        1. Definition
        2. Geometry
oldest in center → 3. Anticline ✓   y-axis symmetry an "∧", 2 limbs l. or right
youngest " → 4. Syncline   same but an "∪", 2 limbs, l. or right
        5. Monocline — only one limb.
        6. Fold orientations (symmetry)   — horiz line — axis
    B. Joints
    C. Faults
        1. Definition
        2. Dip-slip faults
            a. normal
            b. reverse
            c. thrust
        3. Strike-slip faults
            a. right strike-slip
            b. left strike-slip
        4. Oblique-slip faults (both horiz./vertical displacement)
    D. Unconformities
        1. Disconformity
        2. Angular unconformity
        3. Nonconformity

IV. Faults
    A. Dip-Slip (vertical) happens in compression/tension. areas
        1. Normal fault
            >90° Headwall \ Footwall
        2. Reverse Fault
            <45°   Compression zone — something slips.

FORCES PRODUCING STRUCTURAL DEFORMATION

The three primary stress configurations created by interacting segments of the lithosphere are **compression, tension,** and **shearing,** and each produces a characteristic set of deformation (strain) features. These three configurations also represent the three generic types of lithospheric plate boundaries. Specific examples of each are presented in Chapter 8.

Table 6-1

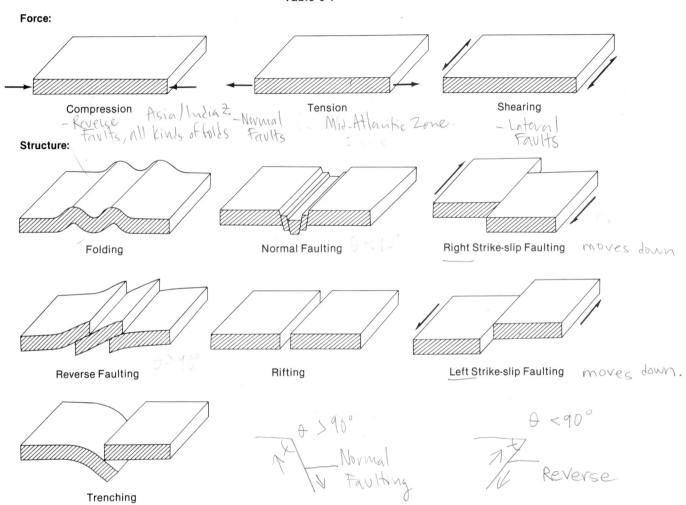

A fourth force involved in structural deformation is the vertical uplifting or subsidence of extensive land areas within a lithospheric segment. This action is a result of lateral pushing, local changes in mantle volume, or surface loading or unloading. Uplift will naturally accelerate erosion and remove of rock material; sinking encourages further deposition. Characteristic structures are:

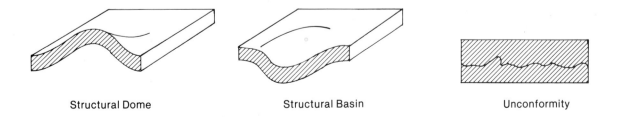

146 Exploring Geology

Keep the causative forces in mind as you learn about the various structures in more detail below. Structures will be easier to understand and deal with if you know their cause. Conversely, you will be able to tell what kinds of forces have been applied to an area by interpreting existing structures.

STRIKE AND DIP

Geologists have devised a notation called **strike and dip** (Fig. 6-1) for uniquely defining the orientation of any plane in space. This is an essential first step toward accurately describing deformed rock bodies, and may be applied to *any planar surface*. Strike and dip, which are always at right angles to each other, orient a tilted plane by relating it to a horizontal surface (Fig. 6-1a).

(Aerial) **Strike:** *The compass direction of a horizontal line on a tilted plane.*
(Cross-S) **Dip:** *The acute angle measured perpendicular to the strike between a tilted plane and the horizontal plane.*

Block diagrams, maps, and cross-sections are used to portray rock bodies in this chapter. The symbol for strike and dip (Fig. 6-1b), which is used on the top of block diagrams (Fig. 6-1c) and maps (Fig. 6-1d), has a strike line in true compass orientation, with a short line at right angles and a number in degrees to indicate the direction and amount of dip.

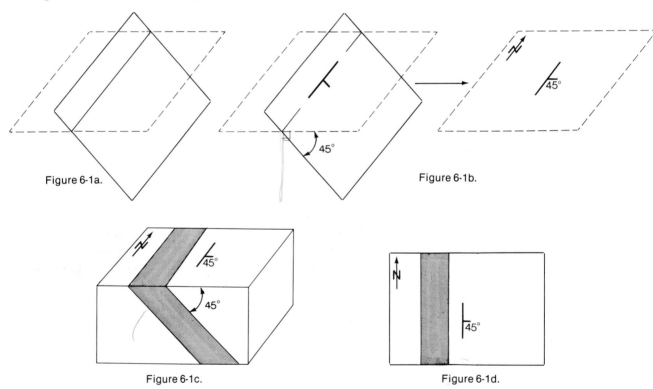

Figure 6-1. Strike and Dip

In the special case of perfectly horizontal beds, the strike is undefined, and the symbol ⊕ is used. At the other extreme, vertical beds are designated ┼ 90° or simply ┼.

Strike and dip may also be conveyed in written or spoken form. For example, the orientation of layers in Figures 6-1c and 6-1d would be stated: N-S, 45°E

Direction of Strike	Amount of Dip	Direction of Dip
N-S	45°	E

The strike line in this case is due north or north-south. Strike orientation may be read from maps by using a protractor. Further examples are provided in Figures 6-1e,f,g.

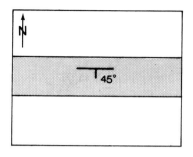

Figure 6-1e. E-W, 45°S

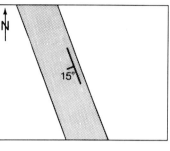

Figure 6-1f. N30°W, 15°W

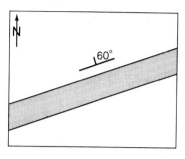

Figure 6.1g. N68°E, 60°N

TYPES OF STRUCTURAL DEFORMATION

Folds

Folds in rock bodies are bends or curves in sedimentary layers, lava flows, or other planar features which demonstrate that the rock body has undergone compressional stress. Rocks which bend under such stress, rather than break, are usually weak (incompetent), or fairly warm, or fluid-filled; or, stress may have been applied slowly.

In order to develop the concepts of fold types and orientation, three elements of fold symmetry should be defined: **axial plane, limb,** and **axis.** An axial plane is a plane which cuts a fold into two symmetrical parts or sides. Each of these sides is called a limb. The axis is a line formed by the intersection of the axial plane and a planar feature in the fold.

There are three common types of folds: **anticlines, synclines,** and **monoclines** (Fig. 6-2).

An *anticline* (Fig. 6-2a) is a fold in which the limbs dip away from the axial plane, and is best defined as a fold in which the oldest beds (lowest numbers) are in the middle. Note the symmetrical distribution of numbers on opposite sides of the axial plane caused by opposite dip directions. Anticlines can serve as traps for gas and oil, and therefore are eagerly sought by petroleum geologists.

A *syncline* (Fig. 6-2b) is a fold in which the limbs dip toward the axial plane, and is best defined as a fold in which the youngest beds (highest numbers) are in the middle. The numbering of beds is symmetrical to the axial plane, but in reverse sequence from those in an anticline. Synclines and anticlines are frequently found in series, alternating with one another.

A *monocline* (Fig. 6-2c) is a local steepening of gently dipping or horizontal layers. It is a fold with one limb, and may be thought of as being one-half of an anticline or a syncline. Monoclines are frequently formed as surface drape of softer, younger layers over more brittle, faulted rocks at depth. This fold/fault relationship will be demonstrated in Chapter 7. Because

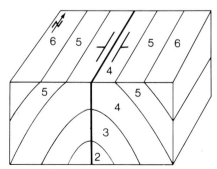

Figure 6-2a. Anticline

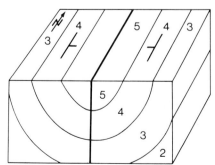

Figure 6-2b. Syncline

Figure 6-2c. Monocline

Figure 6-2. Common Fold Types

148 Exploring Geology

the dip changes in angle but not direction, the numbering of layers across a monocline is continuous rather than a mirror-image sequence.

The *symmetry* of anticlines and synclines depends upon the orientation of the causative stresses in the Earth's crust. The orientation of the resulting folds will affect the pattern a fold makes on a map. Symmetrical, asymmetrical, overturned, and plunging folds are defined below:

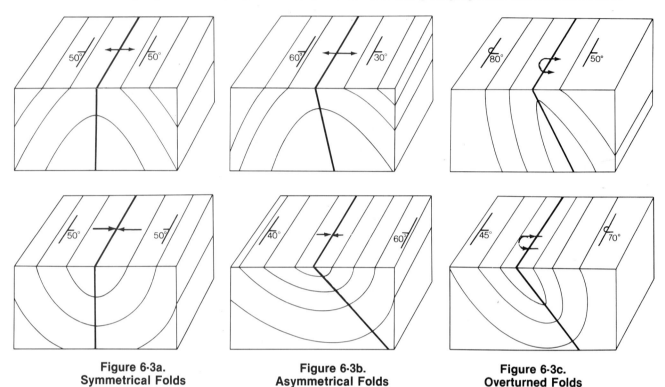

Figure 6-3a. Symmetrical Folds

In a symmetrical fold the axial plane is vertical. The outcrop (map) width of a given layer will be the same on either side of the axial plane trace. Dip angles are essentially the same on both sides of the axis, but opposite in direction.

Figure 6-3b. Asymmetrical Folds

In an asymmetrical fold the axial plane is inclined. Dips on either side of the axis are opposed and unequal. The outcrop width of a given layer will be narrower on the more steeply dipping limb, and wider where more gentle dips prevail (see page 186).

Figure 6-3c. Overturned Folds

In overturned folds, the axial plane is inclined, and both limbs of the fold dip in the same direction. Note that the strike and dip symbol for overturned layers is different from layers which have been tilted less than 90° (see page 183).

If any of the folds shown above (Figs. 6-2, 6-3) were to be tilted so that their axes were inclined to the horizontal, the result would be a **plunging fold**. Plunging fold outcrop patterns are convergent rather than parallel, and are V- or U-shaped on geologic maps. Plunging folds in series are easy to spot by the zig-zag design they form on maps and aerial photos. In cross-section, plunging folds look the same as non-plunging folds.

Anticlines and synclines of any orientation may always be distinguished from one another by the distribution of the ages of layers. This distinction is further clarified for plunging folds by the use of plunge arrows (Figs. 6-4a,b,c), which are arrows on the fold axis pointing in the direction of downward axial plunge. Note that three separate types of information are provided here which precisely distinguish anticlines from synclines: age relationships, plunge arrows, and dip and strike symbols. Figure 6-4d shows a series of plunging folds with appropriate symbols. Anticlines plunge toward the closed end of the V-shaped outcrop, while synclines plunge toward the open end.

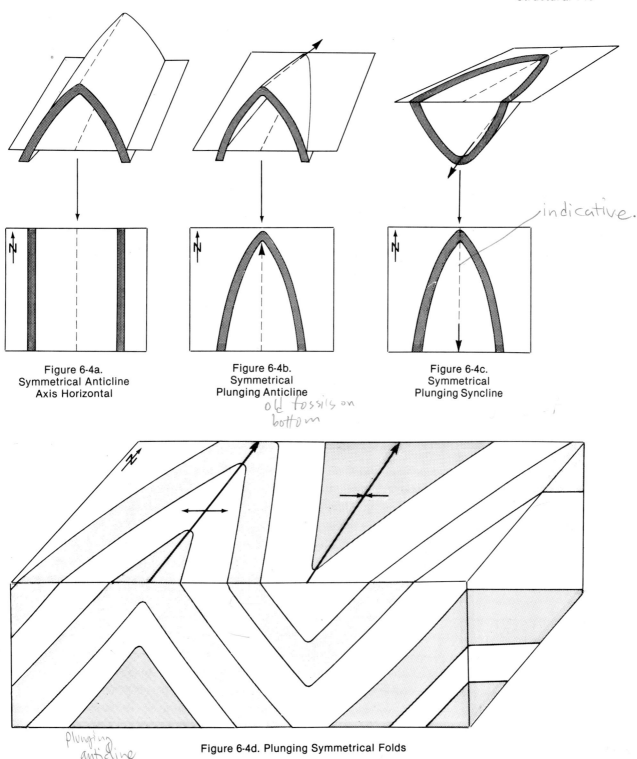

Figure 6-4a. Symmetrical Anticline Axis Horizontal

Figure 6-4b. Symmetrical Plunging Anticline

Figure 6-4c. Symmetrical Plunging Syncline

Figure 6-4d. Plunging Symmetrical Folds

Figure 6-4. Plunging Folds and Map Symbols

Joints

Joints are cracks or fissures in rocks, generally without displacement, caused by overload pressure, shrinkage from cooling, expansion due to erosion of overburden, and various other factors. They occur in all three rock families.

150 Exploring Geology

Faults

A fault is a break in the Earth along which there has been movement of one side relative to the other. The effect of faulting on rocks is to *separate* or *offset* formerly continuous features, such as sedimentary layers, metamorphic foliation, or dikes. It is this offset which often reveals the existence of faults in the field. Another indication of the existence of faulting is the presence of **slickensides**, which are polished surfaces, often with grooves or scratch marks on the fault parallel to the direction of movement. Slickensides are the result of grinding and polishing during fault movement and, where present, may give the best clue to the actual direction of fault movement.

Rocks which break and shift, i.e., fault, under stress are usually relatively strong (competent). They are able to store stress until their ultimate strength is exceeded, rather than bending progressively as stress is applied. Very large faults, with extensive offset and earthquake activity, often constitute the boundary between moving plates. Faults, as planar features, possess strike and dip. The fault trace on a horizontal surface represents the strike, and an arrow attached at right angles shows the dip direction (Fig. 6-5). This arrangement distinguishes the fault symbol from that used for tilted layers, in which the ⊢ symbol is used.

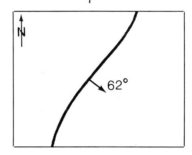

Figure 6-5. Strike and Dip of Fault

Three categories of faults are classified on the basis of direction of relative movement of the rock bodies involved: **dip-slip faults, strike-slip faults,** and **oblique-slip faults**.

Dip-slip faults are those in which displacement is parallel to the dip direction of the fault surface. Rock bodies move up or down the fault surface and slickenside grooves are perpendicular to fault strikes. The three types of dip-slip faults are **normal, reverse,** and **thrust**.

Normal dip-slip faults are those in which the fault surface dips toward the structurally lowered side (Fig. 6-6). Because we cannot know which side of the fault actually moved, displacement is relative. Pairs of arrows are used to express offset in the vertical plane. **U, D** notation (map view) indicates relative *up* or *down* motion on the fault plane. After erosion has removed part or all of the structurally uplifted side, older rocks will be exposed here, while younger rocks are preserved on the down-thrown side (Fig. 6-6d).

Normal faults are frequently the result of extensional forces in the crust, as can be seen by comparing the length of line X-Y before faulting (Fig. 6-6a) to the longer line X'-Y' after faulting (Fig. 6-6b). The effect of normal faulting is to create a partial or total *gap* between the faulted portions of a particular layer. Sets of normal faults in the crust are common over sites of mantle swelling, or where a crustal mass is beginning to split apart. In North America, the Basin and Range Province of the western United States has experienced considerable crustal extension. Crustal blocks downdropped along parallel normal faults are called **graben**, and uplifted blocks are **horsts**. In series, they form *graben and horst topography* (Fig. 6-6d).

A near-surface type of normal displacement is shown by the downhill movement of coherent blocks under the force of gravity (Fig. 6-6e). These slump-type landslides most commonly develop on sparsely vegetated, incompetent rocks (shales, volcanic ash) in slope. Individual landslide events are usually triggered by excess water.

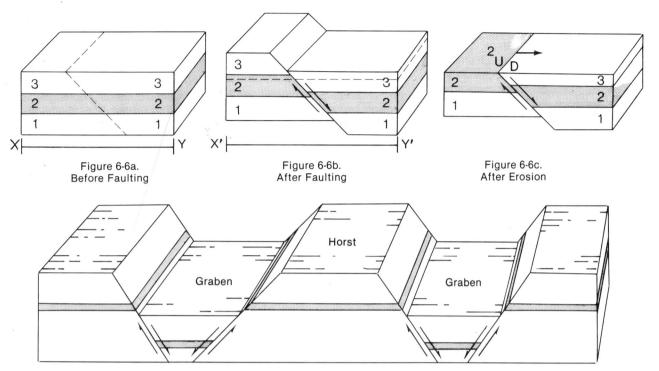

Figure 6-6a. Before Faulting

Figure 6-6b. After Faulting

Figure 6-6c. After Erosion

Figure 6-6d. Graben and Horst Topography

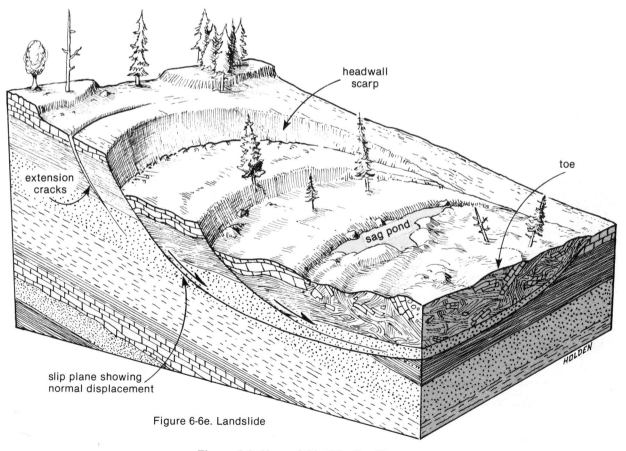

Figure 6-6e. Landslide

Figure 6-6. Normal Dip-Slip Faulting

152 Exploring Geology

Reverse dip-slip faults are those in which the fault plane dips toward the structurally elevated side (Fig. 6-7). They are the result of crustal shortening, or compression. They are genetically akin to folding, in that rock bodies are forced to take up less space as a result of deformation. Note that X'-Y' (Fig. 6-7b) is shorter than X-Y (Fig. 6-7a). Again, as in normal dip-slip faulting, after erosion the oldest bed is on the upthrown side. The difference is that in normal faulting, the dip arrow points to the downthrown side, while with reverse faulting, the arrow points toward the upthrown side.

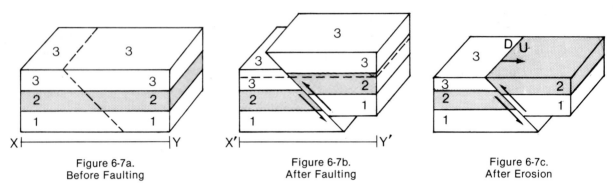

Figure 6-7a. Before Faulting

Figure 6-7b. After Faulting

Figure 6-7c. After Erosion

Figure 6-7. Reverse Dip-Slip Faulting

Thrust faults (Fig. 6-8) are dip slip having low dip angles (<45°). Many thrust faults have significant (tens of kilometers) displacement. With this kind of faulting, movement as shown by the black arrow (Fig. 6-8b) is usually actual, rather than relative, and the transported segment is called an overthrust sheet. On geologic maps the edge of a thrust sheet is shown by a sawtooth line with teeth on the upper plate. Like the uplifted portions of normal and reverse faults, overthrust sheets are subject to erosion. This erosion often leaves isolated older rock remnants called klippen stranded over younger rock sequences (Fig. 6-8c). In areas of extreme crustal compression, such as plate collision boundaries, thrust sheets crawl up on one another; their leading edges tuck under anticlinally from friction (Fig. 6-8d). A hole eroded into the upper sheet of a thrust is called a *window* or *fenster*.

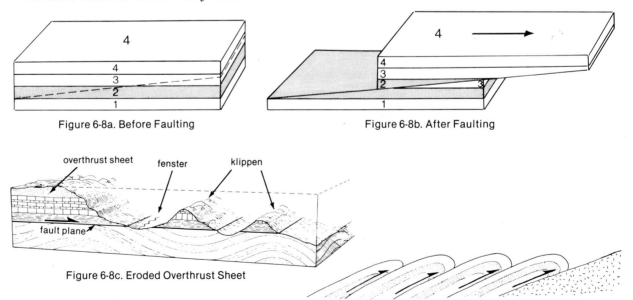

Figure 6-8a. Before Faulting

Figure 6-8b. After Faulting

Figure 6-8c. Eroded Overthrust Sheet

Figure 6-8d. Series of Overthrust Sheets

Figure 6-8. Thrust Faulting.

Strike-slip faults are those in which displacement is parallel to the strike of the fault surface (Figs. 6-9, 6-10). Rock bodies move horizontally (laterally) past one another, and slickensides on the fault surface have horizontal grooves. The two types of strike-slip faults are right slip and left slip, designating the relative movement of one side of the fault past the other. These faults may originate as strain features within a crustal plate, or may constitute the boundary between two plates. Large strike-slip faulting results in greater displacements than large dip-slip faults, in part because movement requires overcoming friction, but not gravity.

Right strike-slip faults (Figs. 6-9a,b,c) are those in which the block across the fault from an observer have moved relatively to the observer's right. Another way to view these faults is that an observer walking toward the fault from either direction along a specific layer or dike would have to go to the right to find the continuation of that layer or dike. Relative movement arrows appear on the map view. **A** and **T** in the cross-section signify movement *away from* and *toward* the observer.

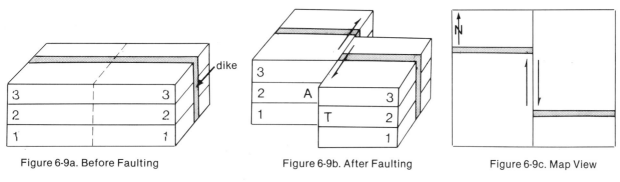

Figure 6-9a. Before Faulting   Figure 6-9b. After Faulting   Figure 6-9c. Map View

Figure 6-9. Right Strike-Slip Faulting

Left strike-slip faults (Fig. 6-10) are those in which the block across the fault has moved to the left relative to the observer, and the continuation of a feature across the fault will be found displaced to the left. Symbols have the same meaning as for right strike-slip faults.

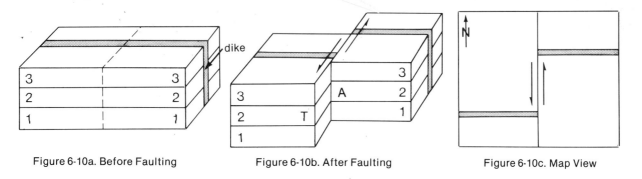

Figure 6-10a. Before Faulting   Figure 6-10b. After Faulting   Figure 6-10c. Map View

Figure 6-10. Left Strike-Slip Faulting

Oblique-slip faults combine both vertical and horizontal offset (Fig. 6-11). This combination is common in nature; however, most faults primarily demonstrate vertical or horizontal slip.

154 Exploring Geology

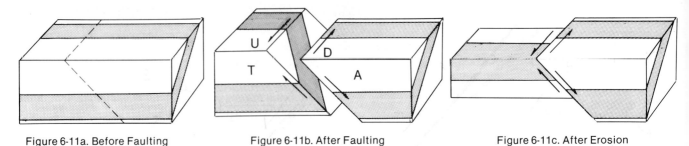

Figure 6-11a. Before Faulting   Figure 6-11b. After Faulting   Figure 6-11c. After Erosion

Figure 6-11. Oblique-Slip Faulting

Unconformities

An *unconformity* is a buried surface of erosion (Fig. 6-12). Because erosion precludes deposition and building-up of time-continuous rock layers, we may also view unconformities as surfaces in the geologic record along which some portion of the rock record is missing. Some types of unconformities are minor, representing only a few tens of thousands of years of erosion, created, for instance, during a temporary drop in sea level. Others, in which layers containing fossils of advanced life forms rest directly upon rocks which are older than any life on Earth, represent profound gaps in the geologic record. The recognition of unconformities is essential to understanding the geologic history of an area.

The three major types of unconformities are classified according to the characteristics of rock bodies above and below the buried surface of erosion, and listed in order of increasing time missing:

Disconformity: Layers below and above the unconformity are parallel; they have the same strike and dip (Fig. 6-12a).

Angular Unconformity: Layers below and above the unconformity are *not* parallel (Fig. 6-12b); they have a different strike and dip.

Nonconformity: The buried erosion surface has been cut on crystalline (plutonic or metamorphic) rocks (Fig. 6-12c).

In general, a disconformity tends to be less profound (have less time missing) than an angular unconformity, and nonconformities usually represent the largest time gaps in the geologic record. Each diagram that follows is accompanied by the simplest possible geologic history that would account for the resulting cross-section. Assume that numbers represent consecutive periods of geologic time, and that sedimentary layers and lava flows were originally horizontal.

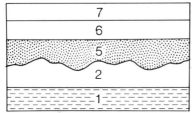 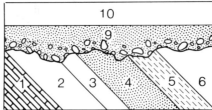 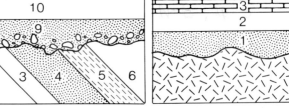

Figure 6-12a. Disconformity
1. Beds 1, 2 deposited.
2. Erosion.
3. Beds 5-7 deposited.

Figure 6-12b. Angular Unconformity
1. Beds 1-6 deposited.
2. Beds 1-6 tilted.
3. Erosion.
4. Beds 9, 10 deposited.

Figure 6-12c. Nonconformity
1. Granite formed.
2. Granite exposed by erosion.
3. Beds 1-3 deposited.

Figure 6-12. Unconformities

Tilted sedimentary layers showing strike and dip, Imperial Valley, California. (Photo by Shannon O'Dunn.)

Symmetrical plunging anticline and syncline, Calico Mountains, Mojave Desert, California. (Photo by Shannon O'Dunn.)

Symmetrical syncline beneath angular unconformity, Barstow, California. (Photo by Shannon O'Dunn.)

Overturned anticline. (Photo by John Shelton.)

Symmetrical plunging anticline. (Photo by John Shelton.)

Asymmetrical plunging anticline. (Photo by John Shelton.)

Syncline plunging to right on skyline. (Photo by Shannon O'Dunn.)

Photo 6-1. Dip and Strike, Folds

156 Exploring Geology

Normal faulting of Columbia Plateau basalt flows, central Oregon. (Grossmont College Photo)

Vertical fault in granite shows slickensides and the non-linear character typical of many faults. San Diego County, California (Photo by Shannon O'Dunn.)

Normal faulting in marine sandstones resulting from large-scale seaward landsliding, Del Mar, California. (Photo by Shannon O'Dunn.)

A tectonic breccia produced by great sliding force at the base of the Titus Canyon Thrust Fault, Death Valley, California. (Photo by Roland Brady.)

Two angular unconformities can be distinguished in these sandstone and conglomerate deposits near Ensenada, Baja California, Mexico. (Photo by Shannon O'Dunn.)

A nonconformity which represents about a billion years of missing Earth history separates crystalline rocks below from gently dipping overlying sandstone. Colorado Rockies. (Photo by Shannon O'Dunn.)

Photo 6-2. Faults, Unconformities

Structural 175

Name_____

Structural Geology Exercise #9: Section _____
Geologic History of the Grand Canyon

The cross-section below represents the rock bodies and structure of the Grand Canyon, Arizona, exposed by downcutting of the Colorado River as the Colorado Plateau has been uplifted.

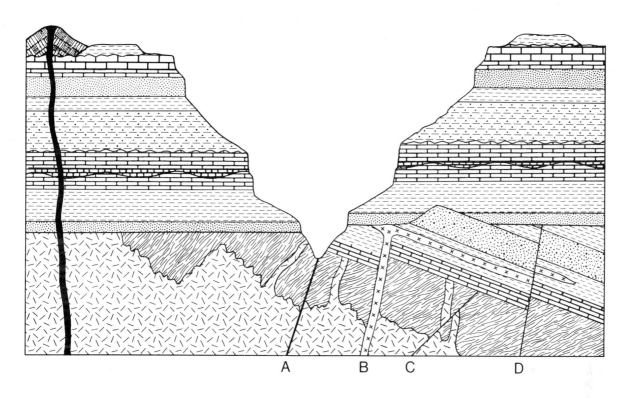

Outline the geologic history of the Grand Canyon on the lines below. Be sure to account for each rock body, and use letters below the cross-section to refer to specific structures.

1. _____ 11. _____

2. _____ 12. _____

3. _____ 13. _____

4. _____ 14. _____

5. _____ 15. _____

6. _____ 16. _____

7. _____ 17. _____

8. _____ 18. _____

9. _____ 19. _____

10. _____ 20. _____

FINAL X-CREDIT NOV. 18 12:30
exact answers only.

Structural 177

Name_____

Chapter 6: Structural Geology   Section _____

REVIEW QUESTIONS

1. What is the fixed angular relationship between dip and strike? _____
_____

2. Anticlines, synclines, and __reverse, Thrust__ faults are caused by compression.

3. What type of fold has the most economic significance to geologists?

_____

4. _____ folds frequently overlie dip-slip faults at depth.

5. _____ folds form U- or V-shaped outcrop patterns on the surface.

6. The best indicators of actual fault movement direction are _____.

7.

Put numbers on the beds to show relative ages. Assume that the land is flat, and that the beds have not been overturned.

8.

Put dip and strike symbols on the beds at the dots. Use the same assumptions as Question #7.

9. What is the difference between a monocline and other folds? _____
_____

178 Exploring Geology

10.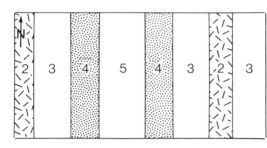

Place plunge arrows on the map, and dip and strike symbols at the dots.

plunges in same direction.

11. What general kind of structure dominates this region?

_____

12. Graben and horst topography is the surface expression of _____ faulting.

13. Complete the block diagram below. Outline the geologic history of this area using specific terms for rock types and structures.

Geologic History

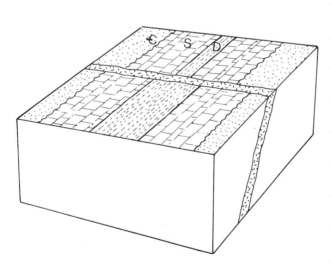

_____
_____
_____
_____
_____
_____
_____
_____
_____
_____

14. How would you distinguish between a disconformity and an angular unconformity on a geologic map? _____

_____

Structural 179

Name _____

Chapter 6: Structural Geology   Section _____

MORE CHALLENGING QUESTIONS

1. What kind of unconformity would be hardest to identity in the field? _____
_____

2. Why do dip and strike symbols appear only in map view? _____
_____
_____

3. An overturned fold is defined as one in which _____
_____
_____

4. What is the difference between stress and strain? _____
_____

5. a. Place the appropriate symbols on the map (top of the block).

   b. Write out the strike and dip for the tilted layers.
   _____

   c. Write out the strike and dip for the fault.
   _____

6. Under compressional stress, some rock bodies bend (fold) and others break (fault). What conditions might determine which strain pattern develops?

   a. Folding _____
   _____

   b. Faulting _____
   _____

180 Exploring Geology

7. Put symbols for rock units in their proper chronological sequence. Place strike and dip symbols on the dots on the map.

youngest

oldest

8. _____ faults have the most displacement, while _____ faults are often responsible for changes in elevation of the Earth's crust.

9. In studying asymmetrical folds, we saw that the outcrop (map) width of a given layer of constant thickness may vary depending upon _____.

Dip of limbs – topography

10. This is a map view of an eroded plain.

a. Assuming horizontal slickenside grooves, name the fault:
  Right Strike-Slip Fault

b. Assuming vertical slickenside grooves, name the fault:
  _____

11. How would you distinguish between a nonconformity and an intrusive contact on a geologic map?

_____
_____
_____
_____

7

Geologic Maps

INTRODUCTION

Geologic maps portray the distribution of discrete rock bodies and geologic structures on a topographic base. Planimetric details such as roads, buildings, and spot elevations are normally included.

Geologists create these informative and often colorful maps by plotting the location and structural orientation (i.e., dip and strike) of bedrock exposures on a topographic base, using data from the field, photo imagery, and subsurface work where available. Boundary lines between distinct rock units are drawn to show how the area would look with the surface cover stripped away, and structural symbols are plotted. Vertical slices showing subsurface distribution of rock bodies and structures, called cross-sections, may be produced from map and field data.

Making a geologic map requires the geologist to define discrete rock bodies (mappable units) in an area, resolve the age sequence of the rocks, identify the structural features and patterns present, and to extrapolate the distribution of rock units and structures where they are obscured by a surface cover of soil, vegetation, water, and man-made features. This process draws upon a geologist's knowledge of petrology, paleontology, structure, topographic maps, and aerial photography interpretation. Similarly, your ability to derive information from geologic maps in this chapter will greatly depend upon your comprehension of material covered in previous chapters.

ROCK UNIT SYMBOLS AND CONTACTS

A discrete mappable rock unit, or **formation**, is defined on the basis of rock type, age, fossil content, and internal structure or texture. Each formation is distinguished by a unique letter symbol, and color or pattern. Many units, such as the Bright Angel Shale and Toroweap Formation in Grand Canyon, have proper names. Others in less studied areas may carry only an age and rock type designation. In either case, the capital letter denotes a time segment from the geologic time chart, and the lower case (small) letters stand for the proper name or rock type. Letter symbols are chosen on the basis of brevity and lack of ambiguity.

Symbol	Rock Unit
Cambrian ← €ba → Bright Angel	Bright Angel Shale
Permian ← P t → Toroweap	Toroweap Formation
Tertiary ← T v → volcanic	Tertiary volcanic rocks

181

182 Exploring Geology

A **contact** is the boundary between adjacent rock units. Contact symbols on maps vary depending upon the certainty of their existence or placement.

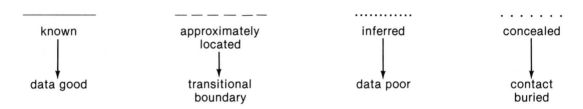

These symbols, plus those for structure and economic prospects, are shown on the opposite page.

Geologic Map 183

GEOLOGIC MAP SYMBOLS
COMMONLY USED ON MAPS OF THE UNITED STATES GEOLOGICAL SURVEY
(Special symbols are shown in explanation)

know 4 sure / *approx.–cover with veget.*

Contact – Dashed where approximately located; short dashed where inferred; dotted where concealed

Contact – Showing dip; well exposed at triangle

Fault – Dashed where approximately located; short dashed where inferred; dotted where concealed

Fault, showing dip – Ball and bar on downthrown side

Normal fault – Hachured on downthrown side

Fault – Showing relative horizontal movement

Thrust fault – Sawteeth on upper plate

Anticline – Showing direction of plunge; dashed where approximately located; dotted where concealed

Asymmetric anticline – Short arrow indicates steeper limb

Overturned anticline – Showing direction of dip of limbs

Syncline – Showing direction of plunge; dashed where approximately located; dotted where concealed

Asymmetric syncline – Short arrow indicates steeper limb

Overturned syncline – Showing direction of dip of limbs

Monocline – Showing direction of plunge of axis

Minor anticline – Showing plunge of axis

Minor syncline – Showing plunge of axis

Strike and dip of beds – Ball indicates top of beds known from sedimentary structures
- Inclined ⊕ Horizontal
- Vertical   Overturned

Strike and dip of foliation
- Inclined ◆ Vertical ✦ Horizontal

Strike and dip of cleavage
- Inclined — Vertical ✦ Horizontal

Bearing and plunge of lineation
- Inclined ● Vertical ↔ Horizontal

Strike and dip of joints
- Inclined ● Vertical ✦ Horizontal

Note: planar symbols (strike and dip of beds, foliation or schistosity, and cleavage) may be combined with linear symbols to record data observed at same locality by superimposed symbols at point of observation. Coexisting planar symbols are shown intersecting at point of observation.

Shafts
- Vertical   Inclined

Adit, tunnel, or slope
- Accessible   Inaccessible

× Prospect

Quarry
- Active   Abandoned

Gravel pit
- × Active   × Abandoned

Oil well
- ○ Drilling   ⌀ Shut-in   ⌀ Dry hole abandoned
- ☼ Gas   ☼ Show of gas
- ● Oil   ✦ Show of oil

STRUCTURE SYMBOLS

Symbols for the structures studied in Chapter 6, with details and variations, are included on the Map Symbol Page. Note that unconformities are contacts between rock bodies, and therefore have no unique symbol. Their existence is verified by age relationships, and frequently by changes in rock type and orientation. Unconformities are often distinguished by wavy lines in cross-section, but not in map view, since a map represents contacts as they appear in the field (without symbols). Deposits of economic significance often have a patterned relationship to specific rock types and structures which may aid in their discovery.

PERIPHERAL INFORMATION ON GEOLOGIC MAPS

Any study of a geologic map should begin with the **Explanation**. All rock units on the map are identified by age, name, symbol, and color. Structure symbols used on the map are also identified.

After studying the map, questions may arise regarding the relationships of rock bodies at depth, and the shape of various structures. Many geologic maps include one or more **cross-sections**, designated **A-A'**, **B-B'**, etc., which portray the third dimension and are very helpful in developing a total picture of the geology in the area mapped.

Occasionally other information is included. Detailed rock unit descriptions, reconstructions of geologically ancient topography, or additional maps emphasizing particular structures, rock units, economic mineral deposits or ground water information may also appear on the sheet. These are included by the geologist who prepared the map to point up a special or important geological feature in the area.

OUTCROP PATTERNS ON GEOLOGIC MAPS

Geologic maps may be visually simple or complex depending upon the number of rock units present, structural complications, and degree of topographical dissection. We may generalize that simple-appearing geologic maps result from a coincidence of topography and structural trends, with few rock units present. Map complexity is maximized where the slope of the land and regional dips differ considerably, and multiple rock units, structures, and dissected slopes are present. Diagrams in Figure 7-1 show some of these variables, and demonstrate how a geologic map compares with the geology in cross-section.

The **map width** of a tabular rock body, such as a sedimentary layer, dike, or lava flow, is controlled by the true thickness of the body and the angle at which the body and land surface meet. Many students are surprised to find that when the meeting angle is small, a thin layer may create a broad band on a geologic map. In Chapter 6 it was shown that on a flat surface, gentle dips produce a relatively wider map width than do steep dips. Consult Figure 7-2 for a graphic presentation of how slope and dip affect map width.

Finally, all tabular rock units, except vertical ones, will show a deflected outcrop pattern where cut by a stream drainage. The **Rule of V's**, showing the relationship between rock unit dips and their outcrop patterns in stream drainages, is shown in Figure 7-3. Understanding this relationship will enable you to determine dip directions on maps where strike and dip symbols are missing.

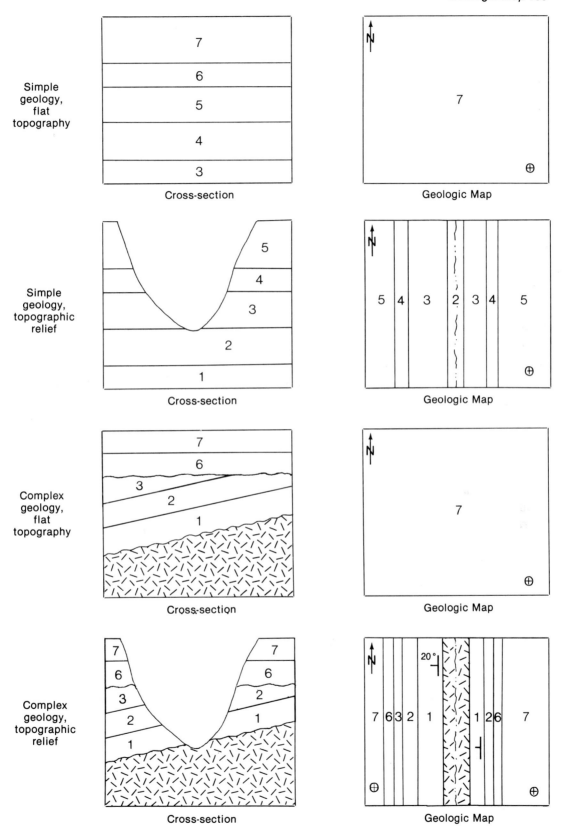

Figure 7-1. Comparison of Geologic Map and Cross-Section

186 Exploring Geology

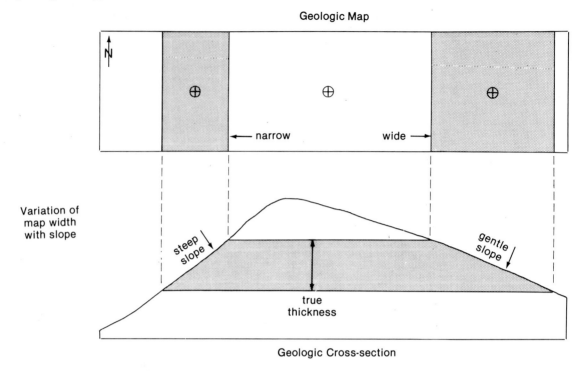

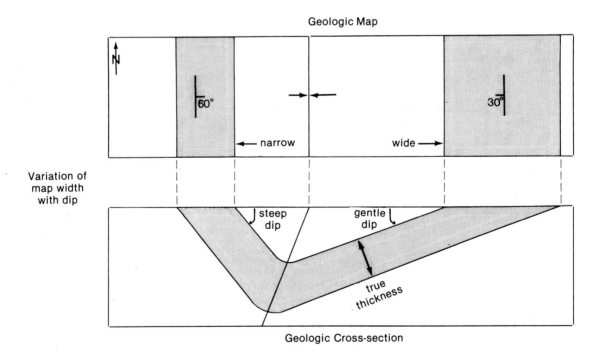

Figure 7-2. Variation of Map Width with Slope and Dip

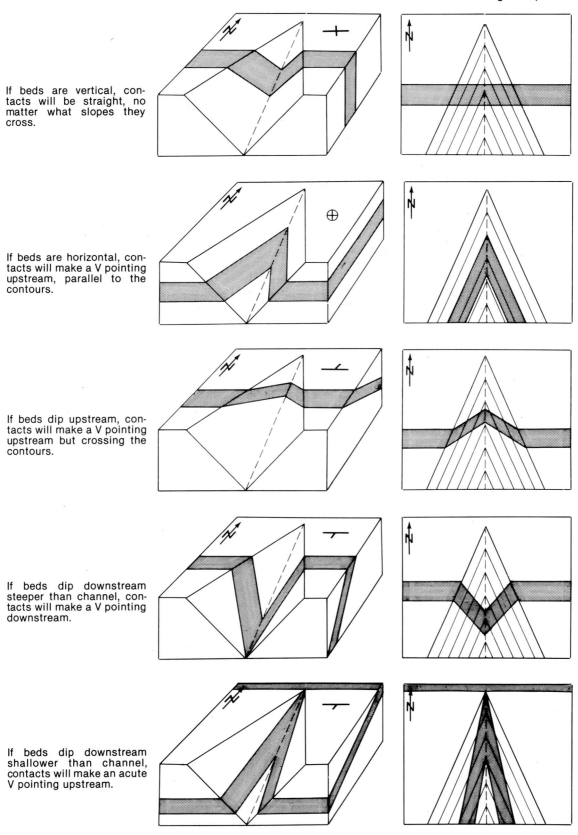

Figure 7-3. Rule of V's

188 Exploring Geology

CONSTRUCTING A GEOLOGIC CROSS-SECTION

Geologic cross-sections are very helpful in understanding geological relationships. If a geologic cross-section is not provided for a given mapped area, it is possible to construct one by following the steps given below. You have already made a number of geologic cross-sections in Chapter 5 by completing front and side portions of block diagrams. The only essential difference here is that we will be combining topographic profiles and more complex geology to produce more sophisticated and realistic cross-sections. Constructing a geologic cross-section is particulary satisfying because you create a view of a geologic setting which did not exist before, and there is room for intelligent speculation as well as straight transfer of data.

1. A cross-section should be oriented as nearly *perpendicular* as possible to major geologic trends such as contacts, fold axes, and fault traces.
2. Construct a topographic profile along the section line, following procedures developed in Chapter 5.
3. Transfer all pertinent geologic data (contacts, fold axes, faults) to the profile by marking them as points along the top edge of the profile paper, and then projecting straight down to the profile line (Fig. 7-4).
4. Where dips are known, extend contact and fault planes down into the profile using a protractor to measure angles.
5. Tie together all outcrops of the same rock unit, using a logical extension of dips, as you did to complete block faces in Chapter 6. Remember that most rock units maintain a fairly uniform thickness, and that units usually do not match up across a fault.
6. Complete the cross-section by placing letter symbols on rock units and displacement symbols on faults.

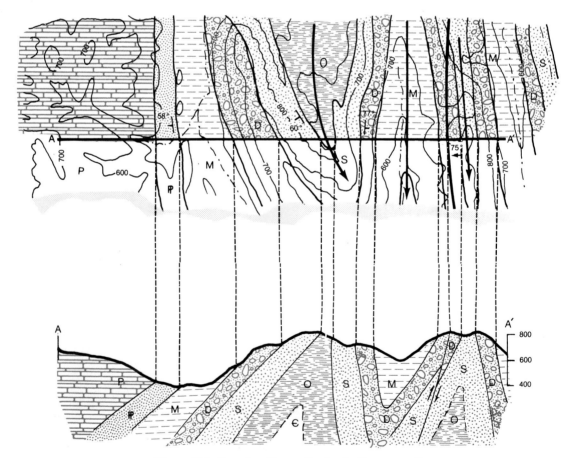

Figure 7-4. Constructing a Geologic Cross-section

Chapter 7: Geologic Maps
Exercises #3 - #5

GEOLOGIC MAP OF THE GRAND CANYON NATIONAL PARK, ARIZONA
Grand Canyon Natural History Association, 1976

MAP LOCATION

Grand Canyon, Arizona
Nearly horizontal sedimentary layers of Paleozoic age form the familiar Grand Canyon panorama. Limestones and sandstones are resistant cliff-formers, while shale units are in slope. This north rim view shows a slight monoclinal folding of the layers.
(Photo by Shannon O'Dunn.)

Introduction to Grand Canyon Geology

The Grand Canyon of the Colorado River is one of the most frequently visited natural splendors of North America. Over a century ago, Major John Wesley Powell called it *"a book of geology."* Virtually every aspect of geology has a representative feature here. An empirical history of this region was developed in Chapter 6 (Ex. #9). This outline can now be filled in with field-derived details of the evolution of this impressive landscape. Opposite page 196 is a central portion of the geologic map showing representative geology of the region. The Explanation and selected cross-sections are included on page 196.

In early Precambrian time the area was dominated by high mountains produced by tectonic uplift. The roots of these long-vanished peaks are represented by the metamorphic and plutonic rocks of the deepest parts of the canyon (lavender and orange on map). Later in the Precambrian, sedimentary rocks were laid down over the eroded roots of the older Precambrian mountains and intruded with diabase (olive, brown, red, and orange north of river). The whole sequence was block faulted, tilted, eroded, and buried by Paleozoic rocks to create the Great Unconformity of the canyon. This complex unconformity separates the steep Inner Gorge from more gentle upper slopes, and older barren rocks from the younger fossil-bearing strata.

The *layer-cake* or well-stratified appearance of the canyon, so obvious from the rim, is created by a nearly horizontal stack of Paleozoic sedimentary rocks (green, lavender, and blue scalloped bands). The area was inundated repeatedly by the sea during the Paleozoic so that marine and non-marine layers alternate, with the latter becoming more common toward the top.

Tertiary uplift and erosion stripped off much of the Mesozoic sequence, and persistent down-cutting during recent geologic times has created the canyon as we see it today. Pleistocene volcanism spilled lava down into the canyon, temporarily damming up the Colorado River. The dams have largely been removed by erosion; however, remnants form some rapids along the modern river course.

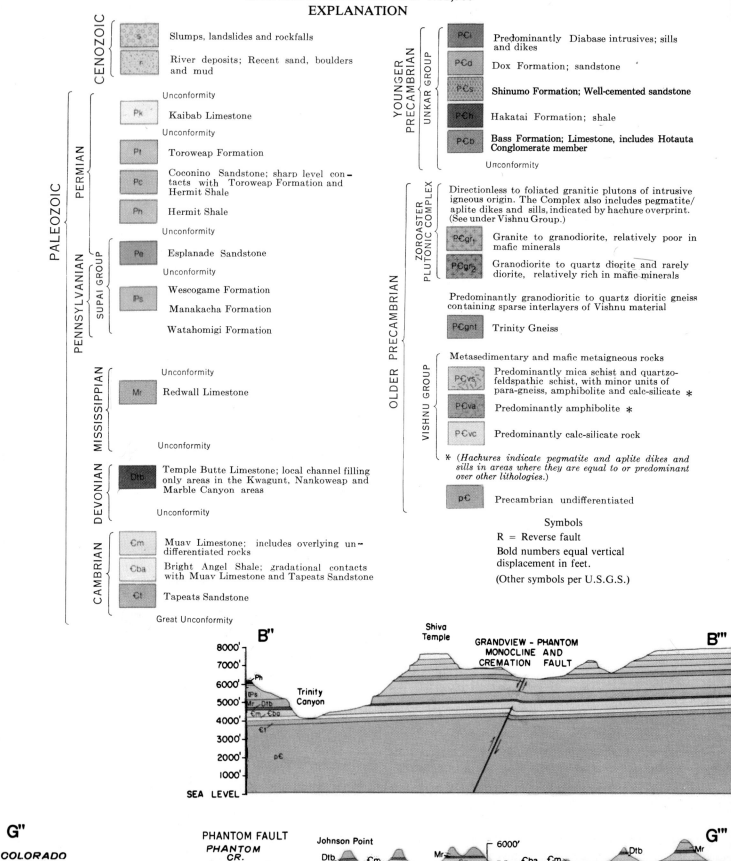

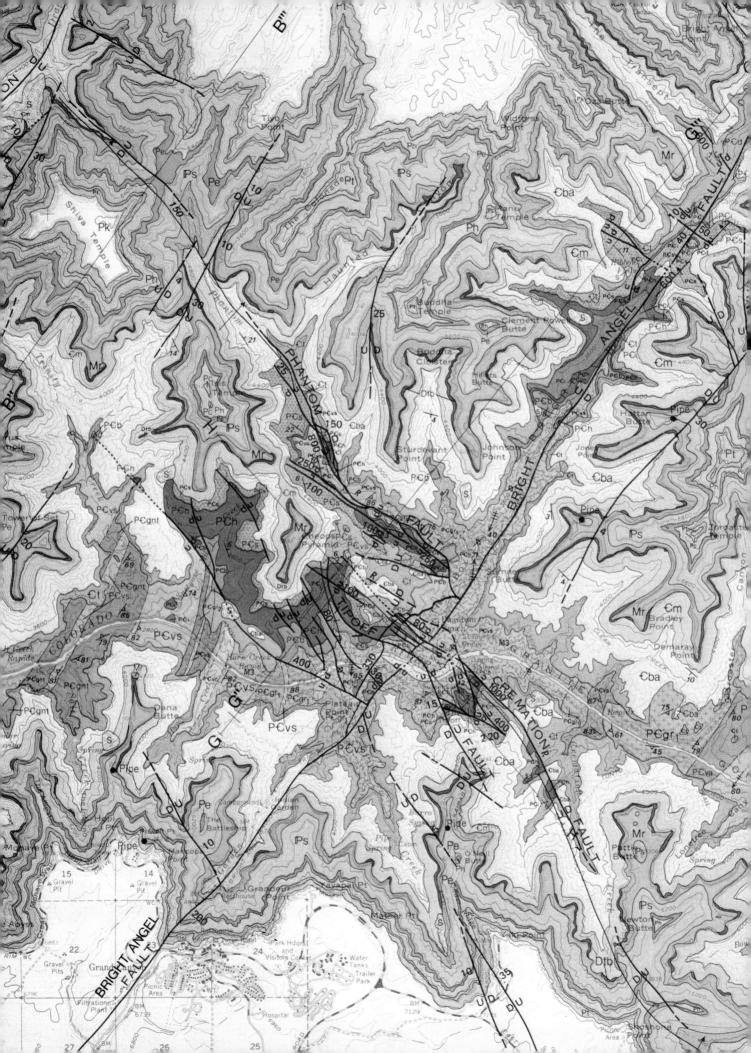

Geologic Maps Exercise #5:
Grand Canyon: More Challenging Questions

Name_____

Section _____

Be sure you have successfully completed Geologic Maps Exercise #4 before going on to these more challenging questions.

1. Why do the contacts of the Paleozoic strata follow the contour lines?

2. Why are dashed contacts shown between pЄva and pЄvs?

3. What rock type(s) might pЄva, Vishnu Amphibolite, originally have been *before* metamorphism?

4. Explain why the North Rim of the canyon is 1,000 feet higher than the South Rim.

5. What kind of tectonic stress has predominated in the area since the beginning of the Paleozoic Era?

6. About 1½ miles south of Isis Temple is a fault which is dashed to the northwest. What kind of fault is it? _____

   What is the age of the fault? _____

7. Why is the Temple Butte Limestone missing between the Muav and Redwall Limestones in so many areas? (Consult *Geologic Maps Exercise #3* and *Map Explanation*.)

8. Why are rapids along the river commonly associated with geologic contacts?

9. Many of the tributary drainages to the Colorado River owe their placement and extent of development to inherent geological weaknesses which encourage stream erosion. What and where are some of these controlling factors?

Geologic Maps 205

Name_____

Section _____

Geologic Maps Exercise #6:

Mapping Contacts from a Photograph

Grand Canyon

Using the map, explanation, cross-sections and Exercise #4, locate the photograph below on the map and study its geologic setting. Then draw in contacts between the different formations, and name each, as accurately as possible. Work lightly in pencil, starting along the left-hand side of the photo where the sequence is unobstructed.

Hints:
1. The top layer is the Kaibab Limestone.
2. The white cliff below the top is the Coconino Sandstone.
3. The river is flowing in schist.
4. Landslide debris is a valid mappable unit.
5. Thin black lines on the photo are intended to delineate fore-, middle-, and background.

Grand Canyon and Colorado River Photo by John Shelton

Geologic Maps Exercise #7
Llano Uplift

GEOLOGIC ATLAS OF TEXAS, LLANO SHEET
Texas Bureau of Economic Geology, 1981

Marble Falls, Texas
Note the diagonal fault in the lower right that separates pre-Cambrian granite to the upper left and Paleozoic shale and limestone to the lower right. In the center is a major granite quarry. The Paleozoic sediments have dropped down (or the pre-Cambrian granite has pushed up) more than 2000 feet. (Photo courtesy Texas Highway Dept.)

The Llano Uplift is a broad, gentle dome in central Texas, in which Precambrian crystalline rocks are exposed. These basement rocks, which are slightly more than one billion years old, consist of gneiss and schist intruded by large granite batholiths.

Resting with profound unconformity on the crystalline basement is a sequence of Lower Paleozoic sedimentary rocks, with an overall maximum thickness of about 2500 feet. These Cambrian and Ordovician rocks are mostly carbonates (limestone and dolomite) with a basal sandstone. They were deposited in a high-salinity environment in which snails and calcium carbonate-secreting algae were the dominant life forms.

Down-faulted areas have preserved a few occurrences of Upper Paleozoic (Pennsylvanian) shallow marine to non-marine sedimentary rocks. Carbonate-secreting invertebrate animals have succeeded the earlier algae as limestone reefbuilders here. Terrestrial plant fossils indicate that at least part of the sequence was deposited above sea level.

From Middle Ordovician to Middle Pennsylvanian time, the land surface was extremely stable—slightly above or slightly below sea level. During the Middle Pennsylvanian, the Llano region was again occupied by a shallow sea in which abundant limestone was formed. This quiet episode was brought to an end by the collision of the North American tectonic plate with another plate that approached from the southeast. The result was the formation of the Ouachita Mountains, which wrapped around the area that was to become the Llano Uplift. At this time the region was broken by dip-slip faults, but it was far enough away from Ouachita mountain-building activity to escape being deformed into folds. However, the region did become elevated above sea level and was a surface of erosion until well into the Cretaceous Period. A broad, shallow seaway invaded not only the Llano area but also most of the low-lying regions of the world.

This episode terminated with renewed upward bulging of the Llano Uplift, which has been above sea level from early Tertiary times to the present. Erosion has removed the Cretaceous cover rocks, which now form an escarpment around the Uplift. In the interior of the Uplift, erosion has revealed the Paleozoic sedimentary rocks lying unconformably upon the Precambrian basement rocks, all of which is broken into large fault blocks.

GEOLOGIC ATLAS OF TEXAS, LLANO SHEET
Texas Bureau of Economic Geology

1:250,000

EXPLANATION

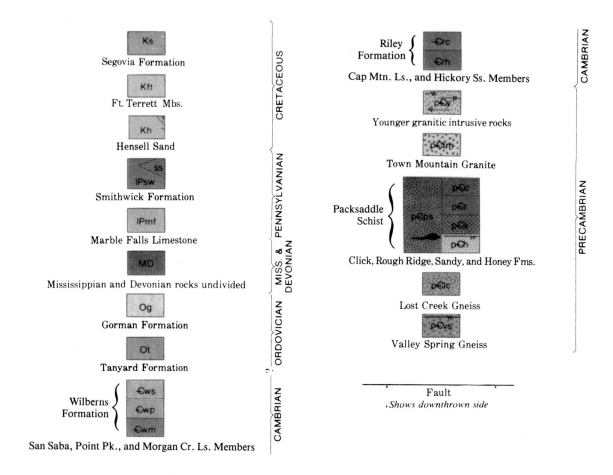

Brief Formational Descriptions

Ks: cherty limestone and dolomite with mollusc fossils and collapse breccia
Kft: cherty limestone and dolomite with mollusc fossils and collapse breccia
Kh: sand, silt, clay, and conglomerate with local cycad (plant) fossils
$\rlap{I}Pcn$: alternating marine units of blue limestone and green shale
$\rlap{I}Psw$: shallow marine to nonmarine limestone, siltstone, shale and sandstone
$\rlap{I}Pmf$: reef limestone with abundant fossils and some dark shale
Og: cherty dolomite and limestone
Ot: cherty dolomite and limestone with gastropods, cephalopods, and trilobites
Cws: cherty dolomite and limestone with stromatolites and trilobites; upper part Ordovician
Cwp: shallow marine siltstone, limestone, and shale
Cwm: marine sandstone and limestone with stromatolites, brachiopods, and trilobites
Crc: fossiliferous marine sandstone and limestone
Crh: sandstone: fine and dirty to coarse and clean with hematite cement
pCy: same as pCtm; mostly in small irregular plutons, dikes, and sills
pCtm: pink, quartz-plagioclase-orthoclase granite; 1.05 billion years old.
pCps: hornblende schist and mica feldspar schist, rare marble and graphite
pCvs, pClc: mostly granite gneiss with minor schist and marble

Geologic Maps 211

Name_____

Geologic Maps Exercise #7:   Section _____
Llano Sheet

After reading the introduction to the Llano region (page 207), take a few minutes to look over the geologic map and accompanying explanation. Then answer the following questions.

1. The Town Mountain Granite (pЄtm) has been exposed to weathering intermittently in the Phanerozoic. Indicate two localities of different ages on the map where a buried ancient soil profile might be located.

    a. _____

    b. _____

2. The lower part of the Hickory Sandstone (Єrh) is an arkose. What does this indicate about the elevation of the Precambrian rocks locally in the Cambrian sea? _____

    _____

3. Which of the crystalline rock units appears to form cylindrical bodies? _____

    _____

4. What is the structural name for the fault-bounded Hickory Sandstone outcrops on the crystalline rocks? _____

5. What is the dip direction of the Cambrian sequence by the McCulloch County line (NE part of map)? _____ What is the dip direction for the same sequence just below the word *Comanche* (north central part of map)? _____

6. The average strike of the dip-slip faults is _____. They must be younger than the _____ rock unit, and older than the _____ rock unit.

7. How can you tell from the map that the Cambrian sequence just east of Panther Creek is essentially flat-lying? _____

    _____

8. Designate the contacts between the following rock units with their correct structural names: e.g., ingeous contact, comfortable sedimentary contact, unconformity (type where possible), fault.

    a. pЄtm (Town Mountain Granite) and pЄvs (Valley Spring Gneiss), eastern edge of map: _____

    b. pЄps (Packsaddle Schist) and Єrh (Hickory Sandstone): _____

    c. Og (Gorman Formation) and Kh (Hensell Sand): _____

212 Exploring Geology

    d. Ot (Tanyard Formation) and Og (Gorman Formation) at top of map: _____

    e. Cretaceous sequence and Cambrian sequence (north central part of map): _____

    f. Cambrian and Ordovician units south of the word *Comanche*: _____

9. Which of the crystalline rock units appears to be the most resistant to weathering? _____

   How can you tell?

10. If you have also studied the Grand Canyon geologic map, give some general parallels between that geologic setting and the Llano Uplift.

Chapter 7: Geological Maps   Name _____

Section _____

REVIEW QUESTIONS

1. Describe an appropriate situation for the use of each of these contacts on a map:

   a. — — — — — —
      approximately located

   b. . . . .
      concealed

2. Why is there no specific or unique map symbol for unconformities? _____
   _____

3. How does the letter symbol for a given rock unit express the age of that unit? Use a specific example. _____

4. Define 'true thickness' and map width. _____
   _____
   _____

5. Explain, using a cross-section sketch, how a flat-lying layer might have a *greater* map width than its true thickness.

6. Explain, using a cross-section sketch, how a flat-lying layer might have a *smaller* map width than its true thickness.

214 Exploring Geology

7. How do map width and true thickness of a vertical layer compare on a horizontal surface?

_____

8. Is there any orientation of a contact or layer which would not show deflection in a stream bed?

_____

9. What is the value of knowing the *Rule of V's* when reading a geologic map?

10. Why is the Grand Canyon considered a good place for geological studies?

Plate Tectonics

8

INTRODUCTION

In 1620 Sir Francis Bacon pointed out that the eastern coast of South America would fit rather nicely into the western coast of Africa. However, it was not until the 1960s that the revolutionary concept of plate tectonics was to offer an explanation for the fit of continents, the occurrence of earthquakes, the origins of ocean basins, and a host of other geologic phenomena. Plate tectonics is a broad-spectrum theory, similar to germ theory in medicine and evolutionary theory in biology, in that it provides a unifying concept to explain a wide variety of apparently unrelated geologic features—from earthquakes in California to the discovery of oil in Nigeria to the topography of ocean floors. The basis of the theory is that the Earth's lithosphere is divided into a number of thin, relatively rigid segments, reminiscent of a cracked eggshell, which are in motion relative to one another. New oceanic crust is formed by oceanic lava welling up at spreading plate boundaries, and older crust disappears back into the mantle as colliding plates plunge one beneath the other. Plate boundaries are marked by a high level of seismic activity, and earthquake epicenters are largely confined to linear belts that mark plate margins. Shallow earthquakes trace the system of ocean ridges, deeper ones mark the plunge of one plate beneath another, and still others mark the collision of plates that form mountain ranges.

As new information has become available the concept of plate tectonics has matured and some "fine tuning" has taken place. Recent research has identified small plates and fragments that have been welded onto the larger plates. These fragments, called *terranes*, are identified by their exotic rock and fossil composition, which does not match that of the area where they are presently located. For example, in northern California and much of Alaska are found rocks typical of oceanic islands, but associated with folded rocks typical of the continent. These terranes became attached to the continent as it overrode oceanic plates being subducted beneath it. Much of southern Europe and the western portion of the United States consists of these "patchwork" bodies of rock, called **accreted terranes**.

What makes plates move? What is the underlying cause? The mantle of the Earth, comprising the area between the rigid lithosphere and the molten core, is a hot solid capable of movement. At the surface of the Earth, the thin rigid lithospheric plates respond in some way to the heat within the mantle. There are two leading theories to explain plate motion. The first states that a rising plume of hot material heated deep within the Earth and confined to a narrow zone reaches the surface along a speading ridge. At the ridge it pushes the plates apart, thus acting as a driving mechanism. The second theory postulates great convection currents within the mantle. Hot material, melted or partially melted, rises to the surface where it cools and sinks back into the mantle to be reabsorbed and recycled. Such convection currents would provide the movement of the plates. Neither of these theories is completely satisfactory, and much research remains to be done before we fully understand the mechanics of plate motion.

216 Exploring Geology

PLATE BOUNDARIES

Plates interact by colliding, pulling apart, or sliding past one another. When plates collide, mountain ranges and volcanos are usually formed. Modern examples of this are the Himalayas and the Andes. If one of the colliding plates consists of ocean floor, which is denser than continental crust, it will plunge beneath the less dense plate and create an oceanic trench such as the Peru-Chile trench west of the Andes. In the ocean, separating plates create new volcanic ocean floor as upwelling magma pushes the plates apart along the ridges that rise from the ocean floor. Within continental areas, diverging plates create rift valleys, which are usually characterized by volcanic activity. Sliding plates are marked by a high level of earthquake activity, as adjacent plates move past one another. The feature produced by this movement is a **transform fault**, which is a strike-slip fault located at a plate boundary or across a spreading center.

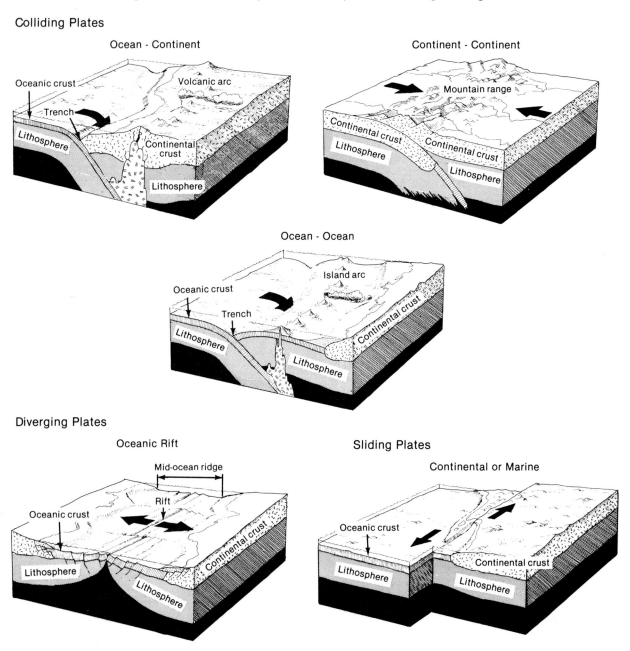

Figure 8-1. Types of Plate Boundaries

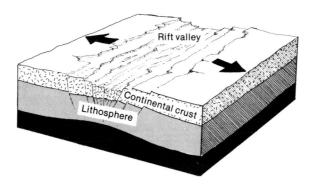

Figure 8-1a. Continental Rift Valley

MANTLE PLUMES AND TRIPLE JUNCTIONS

Areas of high heat flow in the mantle, called "hot spots," cause the formation of magma in the lithosphere. These hot spots are thought to be caused by huge columns of lava welling up from the depths of the Earth in a fixed position under the lithosphere. If the plate above a hot spot changes position, new material is melted and old volcanos become extinct. Over a period of time a chain of volcanos is formed, marking the course of movement of the plate over the hot spot. Some of the island chains in the Pacific are of this type. Dating the volcanic rocks on each of the islands provides an excellent index to the rate of movement of the plate over the hot spot.

Triple junctions are lithospheric localities where three plate boundaries meet. For example, the San Andreas Fault (transform boundary), the Middle America Trench (convergent boundary), and East Pacific Rise (spreading boundary) all converge in the Gulf of California. Another triple junction is located near the Afar Triangle (see Tectonics Exercise #5, p. 230), where the Red Sea, Gulf of Aden, and East African Rift System come together. Here a mantle plume caused the lithosphere to fracture radially into these three spreading boundaries, spaced about 120° apart. As the Red Sea and the Gulf of Aden widen, the Arabian Peninsula is pushed further against Asia. As a result, East Africa is moving away from its parent continent into the Indian Ocean.

Over geologic time, plates have broken up and welded together many times in different patterns. Part of the challenge of modern geology is to reconstruct ancient plate boundaries, discover the hidden seams (called "sutures") of former mountain ranges and shorelines, and then search for the water, petroleum, and mineral reserves that may be hidden in areas where no one would have thought to look before the concept of plate tectonics was formulated.

The study of plate tectonics is complex, and requires the interaction of many scientific disciplines. However, we can interpret the results of past plate movement and determine where past geologic events may have occurred. Based on this information, we can also project future movements, albeit in a time frame difficult to verify. In the following exercises you will be asked both to interpret the past and project the future. Think about as many levels of effect as you can. Consider, for example, how plate movements might affect specific river systems, ocean currents, weather patterns, migration routes of plants and animals, discovery and development of natural resources, and even politics.

It is now believed that the location of many of the major rivers of the world are controlled by plate boundaries. As depressions form, rivers naturally follow them. Where rivers empty into the sea, deltas form. Inland lakes, swamps, and evaporite basins are also associated with rift valleys. All of these features may become major locations for the accumulation of petroleum.

218 Exploring Geology

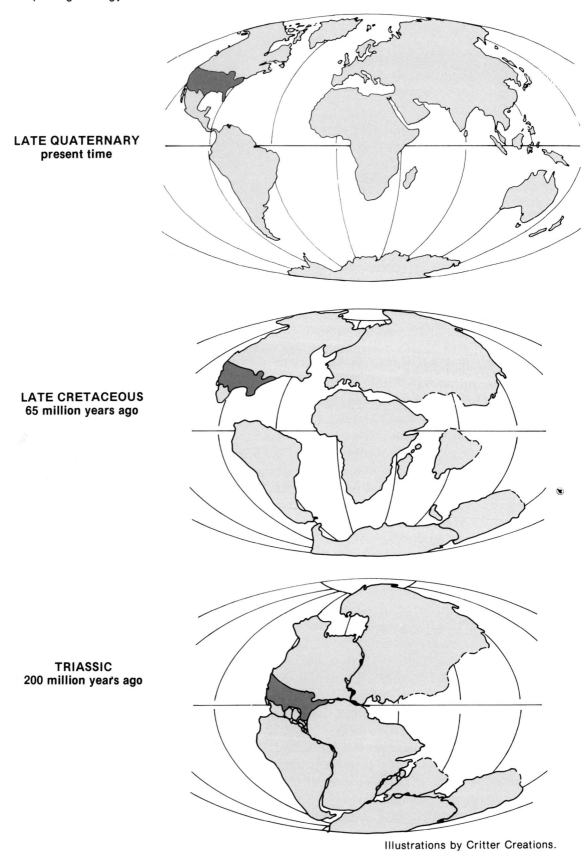

LATE QUATERNARY
present time

LATE CRETACEOUS
65 million years ago

TRIASSIC
200 million years ago

Illustrations by Critter Creations.

Figure 8-2. Continental Positions During Past Historical Periods

Tectonics Exercise #1
World Ocean Physiography

PHYSIOGRAPHIC CHART OF THE
SEA FLOOR
1:106,000,000

Figure 8-3 is a world map which differs from most you have seen. The continental areas are featureless, but ocean basin topography is shown in detail. Because this is a Mercator projection, polar regions are distorted and have been omitted. The purpose of this exercise is to study in detail the orientation and movement of lithosphere plates and the geologic features produced by their interaction.

Your initial assignment will be to locate major plate boundaries and establish relative movement of these plates. In order to do this, review the three types of plate-boundary interactions and the features produced (see pp. 216-17).

> **Diverging Plates:** Where plates separate, mantle material rises up as magma, producing new crust by volcanic activity. **Mid-ocean ridges** are the most common manifestation; where a continent is being pulled apart, large graben or rifts are formed.
>
> **Colliding Plates:** Moving plates inevitably run into one another as they are pushed away from sites of new crustal formation. Oceanic **trenches** mark lines of collision between oceanic and continental masses, and **tectonic mountain chains** may result from the impact of one continental mass against another.
>
> **Plates Sliding Past One Another:** Where adjacent plates are moving tangentially, or in the same general direction but at a different velocity, the feature produced is a **transform fault**, which is a strike-slip fault located at a plate boundary or across a spreading center.

With this information as background, draw in boundaries for the eight plates labeled in Figure 8-3. The rear and leading margins should be distinguished by different colors or symbols. Place motion arrows on each plate to indicate the average direction of travel. Remember that plates move away from ridges where new crust is added, and toward trenches where old crust is destroyed. A few boundaries are obscure but your instructor can help you to locate them.

Exercise questions on page 237 relating to plate tectonics can be answered by using information in the Introduction and by studying Figure 8-3.

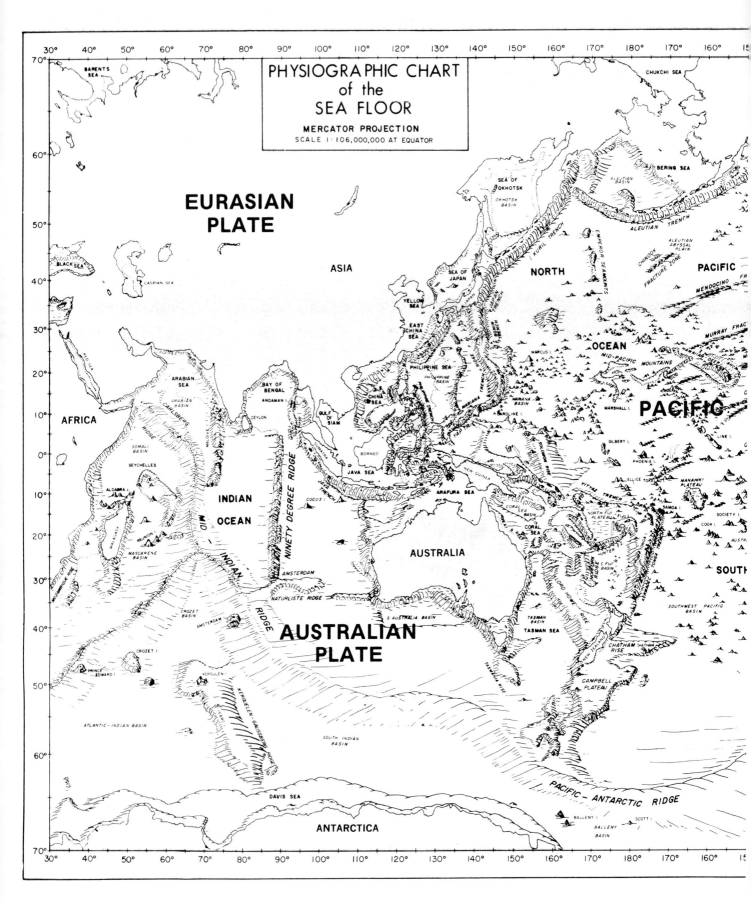

Figure 8-3. Physiographic Chart of the Sea Floor

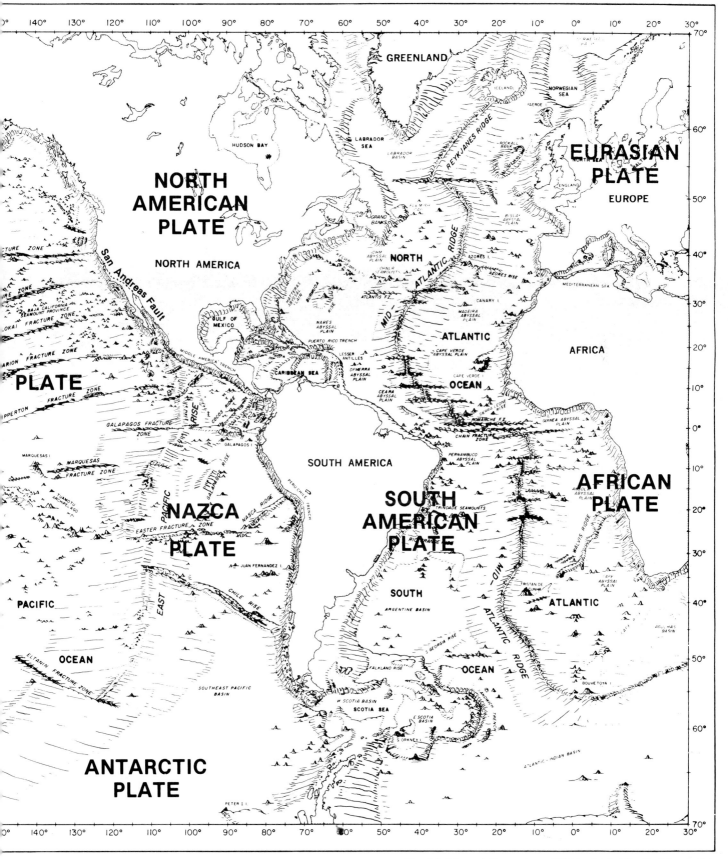

(Reprinted through the courtesy of Hubbard Scientific, Northbrook, IL.)

222 *Exploring Geology*

Tectonics Exercise #2
The North Pacific

THE NORTH PACIFIC 1:46,000,000
OCEAN FLOOR CHART

The chart on the opposite page shows a portion of the North Pacific in considerable detail, including many place names and specific elevations. The Exercise questions on page 239 deal in turn with oceanic features, large-scale western United States tectonic features, and the Hawaiian Island chain.

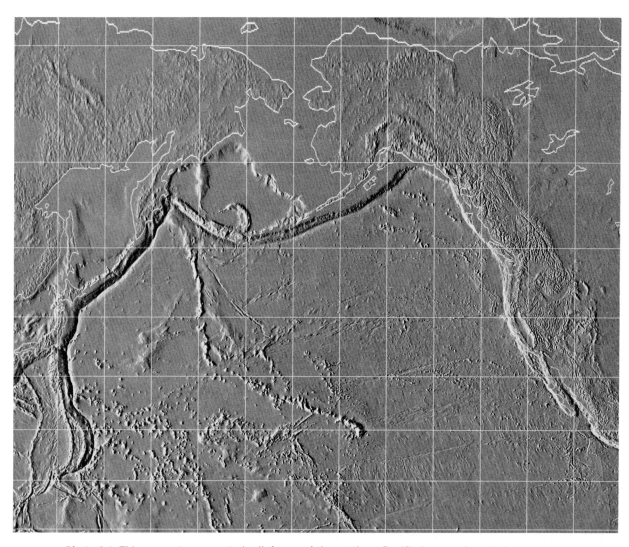

Photo 8-1. This computer-generated relief map of the northern Pacific is part of a relief map of the entire Earth. Original art provided by National Oceanic and Atmosphere Administration. It was published for World Data Center A for Marine Geology and Geophysics by the National Geophysical Data Center, and is part of Report MGG-2, 1985.

Tectonics Exercise #3
A Mantle "Hot Spot" Shield Volcano

KILAUEA CRATER, HAWAII 7½'
U.S.G.S. GEOLOGIC MAP (GQ667)

Kilauea Crater, Hawaii

This is an oblique aerial view of Kilauea Crater looking north. Solidified flows are exposed in the crater wall and floor. The dark irregular area in the middle of the photo is a 1954 lava flow. To the northeast can be seen the subsidence of a portion of the wall. (U.S. Navy Photo.)

This portion of the Kilauea Crater Quadrangle, Hawaii, shows the Kilauea summit caldera (pink, purple, etc.); the uppermost part of the shield volcano (tan); and some flows from the somewhat older shield volcano to the northwest, Mauna Loa (blues, gray).

Kilauea crater, or caldera (see cross-section) was formed in prehistoric time. A caldera is formed by the sudden and voluminous evacuation of a magma chamber at shallow depth. This evacuation removes support from below the eruptive center, and the solid crust collapses into the newly created void. Such collapse features are generally circular, sink hundreds or thousands of feet during the collapse, and eventually fill up with later lavas and ash. Another well-known caldera is occupied by Crater Lake, in the Cascade Mountains of Oregon.

All of the lavas on the map are basaltic. They commonly have a dark gray matrix made up of microscopic crystals of plagioclase, pyroxene, olivine and magnetite with olivine phenocrysts up to one centimeter in diameter. Many of the flows are scoriaceous; pyroclastic materials of the same composition are also present. The flows are of two physical types: glossy, ropy *pahoehoe* and rough, clinkery *aa*.

Kilauea and Mauna Loa, the two active centers of volcanism in the Hawaiian Islands, are welded to three other dormant and extinct volcanos to form the Big Island. Eruptions in the chain, as elsewhere in the oceans, are quiet, due to the low gas and silica content of the magma, and are thus ideal for observation. People in Italy, Japan, and Central America must avoid or flee the aggressive eruptions of their respective volcanos, but tourists and scientists alike flock to the more peaceful events in Hawaii. Exercise questions are on page 241.

KILAUEA CRATER, HAWAII 7½'
U.S.G.S. GEOLOGIC MAP GQ-667
EXPLANATION

MAUNA LOA

KILAUEA

Historic lava
Chiefly consists of products of eruptions between 1823 and 1880

Spatter and lava cones

Keamoku lava flows
kkl, *lobe of Kipuka Kulalio*
kkk, *lobe of Kipuka Kekake*
kl, *lower lobes*

Prehistoric rock units

Lava of Kipuka Pakekake

Lower lavas of Kilauea
Darker shade indicates mappable aa flows

Basalt of Uwekahuna laccolith
Location: Uwekahuna Bluff on northwest wall of Kilauea caldera

Uwekahuna Ash
Location: base of northwest wall of Kilauea caldera

Lower lava of Mauna Loa

QUATERNARY

SYMBOLS

———————
Contact
Dashed where approximately located; dotted where concealed

———•———
Fault
Dashed where approximately located, dotted where concealed. Ball on downthrown side

———
Crack

— — —
Lava channel

Position of edge of Halemaumau before explosive eruption of 1924

Areas of sulfur deposition in 1966

Landslide

Ash redeposited by wind or water

Isopachs of ash and lapilli from Kilauea Iki eruption of 1959, beyond mapped deposits. Contours indicate thickness of 1, 5, and 20 inches (adapted from unpublished map by D. H. Richter)

Approximate outer limit of continuous ash blanket of Keanakakoi Formation of Wentworth (1938). On hachured side of line, thickness of thickest pockets exceeds about 5 feet. The Keanakakoi Formation lies beneath the historic lava flows in the area of the caldera

Tectonics Exercise #4
Composite Volcanos at a Convergent Boundary

THE CASCADE RANGE

The Pacific Ocean rim is traced by a series of arcs where oceanic crust is being subducted under adjacent continents. Partial melting of subducted oceanic crust along with marine sediment becomes magma. Some of this returns to the surface on the continental side of the subduction zones forming a chain of volcanos paralleling the zones. This chain of volcanos, known as the "Pacific Ring of Fire," frequently forms island archipelagos. The Aleutians and the New Hebrides are examples of island archipelagos so formed. Elsewhere volcanos are sometimes perched on continental margins, as in the Andean and Cascade chains.

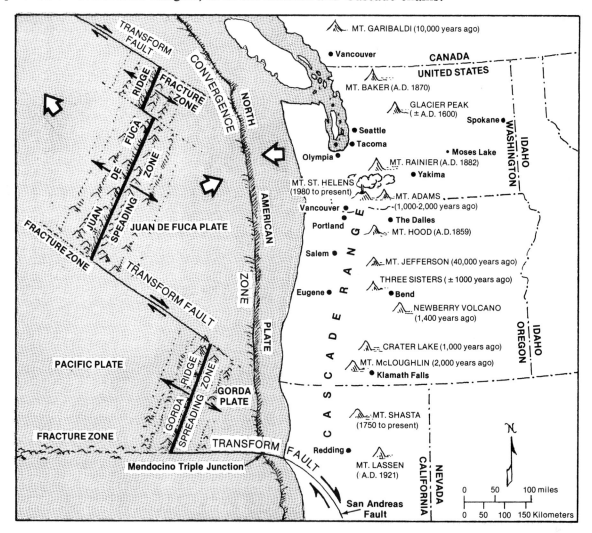

Figure 8-4. This map shows the relationship of the major Cascade volcanos to plate interaction offshore of western North America. Numbers in parentheses are dates of the most recent large scale volcanic activity for each volcano.

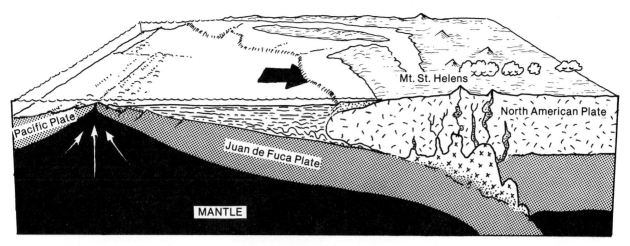

Figure 8-5. This E-W cross-section through Mt. St. Helens shows plate interaction and generation of magma at depth.

The Cascades make up part of the western Cordillera, the active mountain-building portion of North America which experienced reactivated tectonic activity in Pliocene-Holocene time. Cascade volcanos, like others in the Ring of Fire, are of the explosive composite or stratoform type (see page 38). Over a period of time these cones erupt a variety of lava types and textures, ranging from quiet basalt flows to fiery rhyolite clouds. Eruptions may release powerful shock waves, great volumes of ash, and generate tremendous mud flows (see Introduction to Mount St. Helens, page 283).

Cascade volcanic activity began about five million years ago, but the existing cones are all Pleistocene or younger. Magma for these outpourings is being generated by subduction and melting of the Juan de Fuca and Gorda Plates as they move underneath the North American Plate (Fig. 8-5) at the rate of 2 to 3 inches per year. Only two of the fourteen volcanos in the Cascades have erupted in this century—Mt. Lassen in 1914-1921, and Mount St. Helens from 1980 to the present. However, all but four of the volcanos have fumaroles, hot springs, steam and ash emission, and/or historic eruptive events. These volcanos must be considered active and capable of future large-scale eruptions.

230 Exploring Geology

Tectonics Exercise #5
Afar Triangle

AFAR TRIANGLE

Afar Triangle

View of the Red Sea triple junction taken by astronaut Harrison Schmidt from Apollo 17 as it returned to Earth. Note the topography around the junction and compare with the map on page 233. Also, note the Persian Gulf and the Zagros Mountains beyond it.
(NASA Photo)

The Red Sea triple junction is located near the Afar Triangle, a patch of former ocean floor elevated as sea floor spreading opened up the Red Sea and Gulf of Aden, a process which began about 10 million years ago. The third arm of the triple junction is the East African rift valley system, an incipient plate boundary marked by large, well-developed graben formed as the African Continent begins to rift apart.

Many notable geologic features are associated with triple junction development in the Middle East. The Ethiopian Highlands (and thus the Blue Nile) south of the Red Sea and the Yemen Highlands on the north side were formed by initial doming of the triple junction. Mount Kenya and Mount Kilimanjaro are two of the numerous volcanos which mark the path of the East African rift, where mantle-derived magma is leaking to the surface along graben boundary faults. In the Red Sea and Gulf of Aden, seawater penetrates into the young rifts almost to melted rock, picking up metals to create mineral-rich hot brine springs on the new ocean floor.

At the northern end of the Red Sea two fault trenches mark the terminus of the spreading center. The eastern fault extends from the Gulf of Aqaba northward through the Dead Sea and the Sea of Galilee (see Stereophoto 8-1). This fracture is usually interpreted as a transform fault that separates the small Arabian plate from the African continent. In this case, the Sinai Peninsula and Israel would be considered stationary while the Arabian side has moved nearly 65 miles to the north. Volcanos and lava flows are common along rift valleys, and new lakes and rivers form due to the changes in topography and drainage.

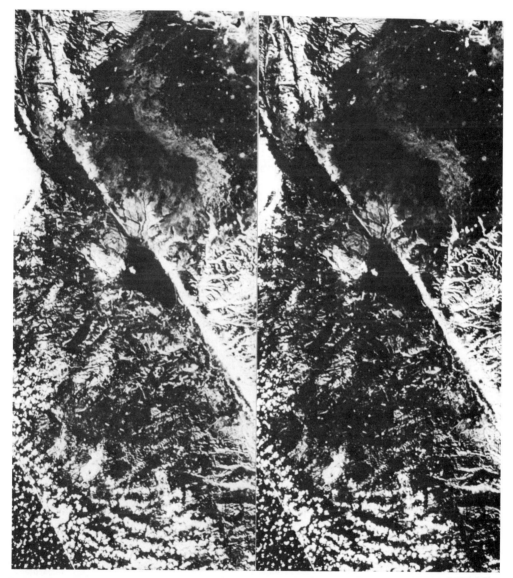

Stereophoto 8-1. View of the Red Sea Rift as it passes through the Sea of Galilee. This fault may be an incipient transform fault from the Red Sea spreading center. Note the smooth volcanic rocks that created the Sea of Galilee by damming the Jordan River.

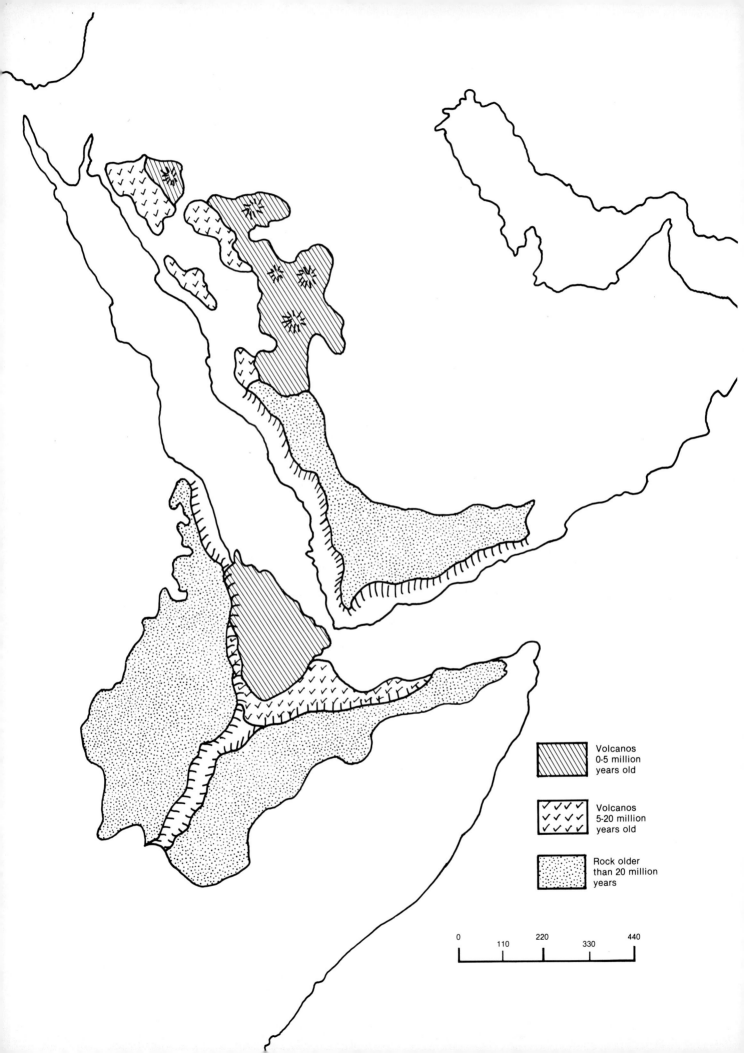

234 Exploring Geology

Tectonics Exercise #6
A Transform Plate Boundary

THE SAN ANDREAS FAULT

San Andreas Fault, CA

The San Andreas fault between Bakersfield and San Luis Obispo shows the classic features of a land-based transform fault. Linear structures usually indicate faulting, as do displaced drainages, escarpments, sag ponds and lakes, and rock units that are out of place. Here the direction of plate motion along the fault is obvious. Major earthquakes have produced displacement of up to 15 feet horizontally along the fault plane.
(U.S.G.S. photo)

The San Andreas Fault is the tectonic boundary between the Pacific Plate on the west and the North American Plate on the east. It extends more than 800 miles, from the Gulf of California (Sea of Cortez) to north of San Francisco. It is not a knife-like cut in the Earth's surface but rather a complex zone of broken rocks and accessory faults. It originated 15-20 million years ago and the Pacific side has moved at least 350 miles to the northwest during that time.

Currently, average movement along the fault is approximately two inches per year. Where smooth, gentle movement (called "creep") does not occur, stress builds up in the rocks and is released in the form of massive earthquakes. Prediction of earthquakes in California is based to a large degree on calculating the build-up of stress in "locked" rocks and trying to estimate when they will break loose. Most geologists interpret the San Andreas as a transform fault that links the East Pacific spreading ridge with its remnant off the coast of northern California at the Mendocino Fracture Zone. It is possible that the Mendocino zone may represent a new triple junction.

The following high altitude stereophotos of the San Francisco area show part of the San Andreas fault system. Mark on the photo the faults as evidenced by linear features. Look for other indications of movement, such as displaced streams and other geological features that do not match across the fault zone. Refer to the SLAR photo of the same areas on page 66.

Stereophoto 8-2. High altitude stereophotos of the San Francisco Peninsula, California. (Photos by NASA.)

Tectonics Exercise #1:
World Ocean Physiography

Name _____

Section _____

QUESTIONS

1. What do the Red Sea and the Gulf of California have in common? What will happen to these areas in the next few million years?

2. Fracture zones (transform faults) in the Atlantic and South Pacific are clearly associated with oceanic ridge systems. Assuming this is also for the Mendocino and other northeast Pacific fractures, where is the ridge? What has happened to it?

3. The Himalaya Mountains form the northern border of India. They contain well-developed compression features, such as reverse and thrust faults, and are rapidly uplifting at the present time. How might their existence be linked to plate motion?

4. Where would you expect to find the oldest ocean-floor rocks on the Pacific Plate? Why?

5. Along the Peru-Chile trench, oceanic crust and continent-derived sediments are subducted under South America and melted to magma, which rises to the surface to form volcanic deposits. What is the average volcanic rock type of these deposits? _____ What type of volcano would be formed (see Table 2-2, Chapter 2)?_____

6. Iceland is an above-water portion of the Mid-Atlantic Ridge consisting of oceanic crust recently derived from the mantle. What is its specific rock type?_____

238 Exploring Geology

7. As ocean crust is produced at a ridge and then moves away, it is a collecting site for sediment raining down from the plankton layer. Where on the North American Plate would you expect to find the thickest accumulation of submerged marine sediments?

8. Continental shelf areas are generally considered to be important sites for petroleum formation. List several countries or geographic areas which may have significant petroleum reserves of this kind.

9. It has been estimated that plate-motion rates are *about* 1½ inches per year. How long ago did the Atlantic begin to open? (Assume that the rate of plate motion has been constant.)

Tectonics 239

Tectonics Exercise #2:
The North Pacific

Name_____

Section _____

1. Most of the ocean features found world-wide have representatives in the North Pacific. For each of the following, give a specific example with elevation or depth, where possible:

   a. Abyssal plain _____

   b. Broad continental shelf _____

   c. Fracture zone _____

   d. Seamount (drowned volcanic peak) _____

   e. Guyot (flat-topped seamount) _____

   f. Site of new crust production _____

   g. Site of crustal destruction _____

   h. Sliding boundary between two plates _____

   i. Composite volcanic chain _____

   j. Shield volcanic chain _____

2. The geology of the western United States is complex and still not completely understood. We know that strike-slip faulting, vertical shifting of the crust, and volcanic activity are taking place at the present time, and that much of this tectonic activity is directly related to plate movement. The following questions encourage you to relate what you have learned about western United States geology to plate tectonics, using illustrations on pages 220-21 and 223.

   a. Briefly describe the dominant geologic process which has created the Gulf of California.

   b. What is the proper and specific structural name for the boundary between the North American Plate and the Pacific Plate?

   c. The San Andreas Fault is offset at Santa Barbara, just north of Los Angeles. What offshore structure aligns with this offset?_____

   What is the specific kind of displacement along this structure? _____

   d. Spreading sites which extend under continental areas create graben-like features called rift valleys, some of which have very low elevations. What surface feature(s) might represent the East Pacific Rise under western North America?

e. The Cascade volcanos have been built up of magma produced where the Pacific and North American plates interact. What average lava composition would you expect in these volcanos?

_____

What type of volcanos would form here? _____

What other volcanic chains around the Pacific share these characteristics? (name three)

_____   _____   _____

f. What plate is Los Angeles on? _____ The Sierra Nevada?

_____

3. More than 100 localized, relatively stationary **plumes** or *"hot spots"* in the upper mantle are scattered around the globe. Although most volcanism takes place at plate boundaries, these isolated spots account for some spectacular intraplate volcanic features. The Hawaiian Islands exemplify the "hot spot" phenomenon, where a chain of volcanos is created by the progressive movement of a plate over one of these sites of magma production.

   a. The relative age of individual islands may be estimated by the degree to which wave erosion has diminished them subsequent to volcanic construction. Therefore, the Hawaiian Islands increase in

   age toward the _____ (compass direction).

   b. This age progression would confirm that the plume is now located _____,

   and that the Pacific Plate has been moving in a _____ direction.

   c. The oldest island in the chain has been dated at 40 million years. Using the chart on page 223, measure the distance between the oldest and youngest islands (1 inch equals 725 miles), and then compute the rate of plate motion in inches per year.

   d. The Emperor Seamounts continue the volcanic age progression to 80 million years at the Aleutian trench. Assuming that they, too, were formed over the same hot spot, in what direction was the

   Pacific Plate moving between approximately 40 and 80 million years ago? _____
   Since plates move perpendicular to ridge axes, what was the orientation of the operating spread-

   ing ridge during this time? _____ What do these facts

   tell you about the position of ridges and direction of plate motion? _____

   _____

4. From the map of the Hawaiian Islands and the Emperor Seamounts calculate the rate of movement of the Pacific for:

   a. the last 10 million years _____

   b. the period from 10 to 40 million years _____

   c. the period from 40 to 75 million years _____

5. Explain the change in orientation of the volcanos approximately 40 million years ago.

Tectonics 241

Name _____

Tectonics Exercise #3:
Kilauea—A Mantle "Hot Spot" Shield Volcano

Section _____

QUESTIONS

1. How do the ages of the flows from Mauna Loa and Kilauea, respectively, fit in with the developmental history of the Hawaiian Chain as proposed earlier in the chapter?

2. Name the two roughly circular collapse features, or *ring structures*, on this map.
   _____

3. Give the location, proper name, and type (shape) of the one *intrusive* body shown on the map.

4. Where are the 1954 flows?

5. What kind of faulting forms the caldera boundary? _____

6. The red-dashed contours, **isopachs**, indicate the thickness in inches of ash thrown out of a vent east of Kilauea in 1959. From which direction was the wind blowing during that eruption?
   _____

7. What is the structural name of the linear fault-bounded features between the west side of Halemaumau and the Crater Rim road?

8. During what eruption was the Halemaumau fire pit developed to its present-day size?

9. What is the ultimate source for the Hawaiian lavas? _____

10. Based on the historical dates given on the map for the lava flows in the adjacent to Kilauea Crater, eruptions appear to be periodic, coming in groups of years separated by gaps of inactivity. Note these gaps between sequences of eruptions, and on that basis make a prediction of when the next eruptions might be anticipated.

Tectonics Exercise #4:
Composite Volcanos at a Convergent Boundary

Name_____

Section _____

QUESTIONS

1. What plate boundary types converge at the Mendocino triple junction?

   _____

   _____

2. The Cascade Range is one segment of a larger volcanic chain called _____

   _____. It is also part of the active mountain-building segment of North

   America called _____.

3. The most common volcanic rock type of the Cascades is _____.

   This intermediate lava type is created by magmatic mixing of _____

   and _____ in the upper mantle.

4. Why do the Cascades terminate in northern California? (Hint: Look at the plate boundaries.) _____

   _____

   _____

5. Where on the Juan de Fuca Plate would the oldest basalts be found?

   _____

   Explain: _____

   _____

6. What geographic portions of the Cascades have shown the most recent activity?

   _____ and _____.

7. Why are composite chains like the Cascades located landward of the boundary where the plates converge? _____

   _____

   _____

8. Explosive volcanic activity is associated with _____ lava composition, and quiet

   flows with _____ composition.

9. What two population centers are near enough to threatening Cascade volcanos to consider volcanic

   hazards in their overall civil defense plan? _____ and _____.

Tectonics Exercise #5:
Afar Triangle

Name_____

Section _____

QUESTIONS

1. From the information on the attached map, calculate the position of the Arabian Peninsula 5 and 10 million years from now. Draw the two outlines over the map on page 232.

2. Explain the relationship between tectonic patterns in the Middle East and the occurrence of oil in Saudi Arabia.

3. If the Red Sea had opened up 20 million years ago (that is, if it were twice its current age) what differences might exist in Middle Eastern geography and politics?

4. In what direction is the African plate moving? How will its movement affect the Mediterranean Sea and southern Europe now and in the distant future?

5. If East Africa continues its current tectonic activity, what will become of the rift system?

6. How are metallic ore deposits related to sea floor spreading?

7. How does the Afar Triple Junction differ from the triple junction in the Gulf of California?

Tectonics 247

Name_____

Tectonics Exercise #6: Section _____
San Andreas Fault—A Transform Plate Boundary

QUESTIONS

1. List topographic features that show the presence of the San Andreas Fault, such as displaced drainage and linear topography, and locate them on the photos. You should find at least five features related to the fault. Compare with the low angle photo on page 234 and with the SLAR photo on page 66. List all of the fault-related features.

Feature	*Location (or photo)*
_____	_____
_____	_____
_____	_____
_____	_____
_____	_____
_____	_____

2. If the San Andreas Fault continues its current motion for the next 20 million years where will Los Angeles be located? Take into account the motion of the North American Plate and the Pacific Plate. Give the appropriate geographic reference points.

3. The Mendocino Fracture Zone off the coast of California is an apparent triple junction. If so, predict what new plates may form, and their movement over the next 30 million years.

4. What evidence would you look for to determine how far the San Andreas Fault has already moved?

5. What geologic event might cause the San Andreas Fault to stop movement and become inactive?

Chapter 8: Plate Tectonics Name_____

Section _____

REVIEW QUESTIONS

1. Explain the plate tectonic process that produced the following geologic features: a. the Andes Mountains, b. the Ural Mountains, c. the Aleutian Islands, d. Iceland, e. the Himalaya Mountains, f. the Gulf of California.

2. What features would you look for to determine ancient plate boundaries on continents?

3. Explain the origin of oceanic crust and identify its relative age at different locations within a plate.

4. Why are volcanic eruptions in Hawaii relatively non-violent whereas those in the Pacific Northwest are explosive?

5. Microplates which have been welded onto larger plates are called _____.

   and many have been identified in the _____ region of the United States.

6. What is the fundamental difference between driving mechanisms as postulated in the two leading theories of plate motion?

7. Where are the two most extensive zones of plate collision on Earth located?

   _____   _____

8. Earthquakes, which are common at all three types of plate boundaries, are weakest and most shallow along spreading boundaries. Why?

9. In early Mesozoic time, the eastern United States abutted _____.

   What was adjacent to the western United States? _____.

9

Applied Geology

INTRODUCTION

Rocks are the basis of most economic considerations. Geologic conditions affect the value and safety of your home; the quality of water you drink; the price you pay for fuel, electricity, food and textiles; and may determine where you live. In this chapter you will see some of the applications of geology to daily life, and also how it may be brought to bear on more extraordinary, cataclysmic natural events.

The majority of Earth's inhabitants have been affected by the variability of petroleum prices in the last decade or so, as gas-service lines and soaring prices were replaced by the oil glut of the mid-1980s. What determines the price of fossil fuel? Oil is inexpensive to produce but very costly to find. Also, current technology permits recovery of only about 50% of the oil discovered; the rest is still in the ground when the well is pumped "dry." Future oil prices will depend largely on advances in methods of exploration and recovery and on the development and availability of alternate energy sources. Exercises #1 and #2 deal with the geological aspects of petroleum origin, exploration and production, and #1 touches on aspects of coal and gypsum production as well.

Water is the geologic commodity most essential to human life. Exercise #3 will introduce you to the occurrence and pollution hazards of ground water and illustrate some of the problems of dependable production from underground reservoirs.

Many geologists pursue professional studies of an ecological nature, monitoring the well-being of surface waters, animal breeding grounds, and ecologically sensitive locales such as estuaries, the tundra, and deserts. Exercise #4 considers the drastic alteration of a coastal area by humans and the resulting benefits and disadvantages of such alteration.

The last three exercises incorporate examples of the more catastrophic types of geologic activity: those which occur suddenly and often result in loss of property and human life. Exercise #5 introduces a classic urban landslide, one which has caused heavy property damage and also has been extensively studied. In Exercise #6 the student will learn to locate an earthquake's epicenter (surface projection) and something about how earthquakes are measured. Finally, Exercise #7 brings this book to a close where we started on the cover—the May, 1980 eruption of Mount St. Helens composite volcano with its devasting effects on human and animal inhabitants, vegetation, soil, water and air.

Chapter 9 introduces no new theories or geological principles. You will be able to complete the activities successfully by applying what you have learned in earlier chapters to each specific problem or setting detailed here.

Applied Geology Exercise #1
Devil's Tooth Area, Wyoming

ECONOMIC DEPOSITS IN THE STABLE CENTRAL ROCKIES

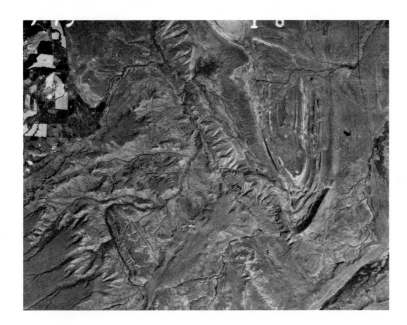

QUADRANGLE LOCATION

Devil's Tooth, Wyoming

This plunging anticline is typical of structures in the Central Rockies. Quiet deposition of Paleozoic and Mesozoic sedimentary rocks was followed by mild folding and faulting in the Tertiary Period. Fossil fuel and gypsum deposits are associated with the formations and structures here.
(U.S.G.S. Photo)

The North American Cordilleran tectonic zone stretches from Alaska through Mexico, and from the Pacific margin to the eastern edge of the Rocky Mountains. Within the zone, western mountain ranges of crystalline rocks and active volcanos, created during Mesozoic and Cenozoic tectonic episodes, abut a central band of plateaus and basins with Cretaceous-Holocene volcanic and sedimentary fill. The eastern Cordillera in the United States is formed by the Rocky Mountains, including the Central Rockies of Wyoming and Colorado.

Forming a buffer zone between the more intensely deformed Cordillera to the west and the stable continental platform to the east, these low-relief mountains are characterized by Paleozoic and Mesozoic sedimentary rocks overlying the Precambrian crystalline core of the continent. These sedimentary units are impressively uniform in thickness and distribution due to the structural stability of this region during the time they were being deposited.

Lower and Middle Paleozoic sandstone, shale and limestone were laid down in a shallow marine environment which experienced mild fluctuations of sea level. As the land emerged above sea level in late Paleozoic to middle Mesozoic time, marine flooding and shale deposition were replaced by terrestrial deposits. Cretaceous marine flooding and shale deposition were succeeded by an influx of late Cretaceous terrestrial deposits, as clastic debris washed off from the rising Cordillera to the west pushed the marine shoreline progressively eastward. Bentonite beds in the Cretaceous sequence derive from volcanic ash from eruptions to the west, which accompanied the emplacement of granitic batholiths. During Tertiary time, this region experienced mild folding, thrust-faulting and uplift. Cross-section A''—A''' on page 254 shows the nature and distribution of some of these structures.

Geological deposits of economic significance in this part of the Rocky Mountains include petroleum, coal, and gypsum. Here, as in the midcontinent and Appalachian regions, Carboniferous (Mississippian and Pennsylvanian) units are often oil-bearing; low-sulfur coal underlies more than 300,000 square miles of the Rocky Mountains. Warm, shallow waters of the Middle Jurassic Sundance Sea deposited locally thick gypsum layers.

DEVIL'S TOOTH, WYOMING 15'
U.S.G.S. GEOLOGIC MAP GQ-817
EXPLANATION

QUATERNARY

Qal — Alluvium
Unconsolidated deposits of silt, sand, gravel, and cobbles along stream valley and at or near present stream level.

Qtc | Qtps
Qtp | Qtr — Terrace gravel
Unconsolidated deposits of gravel, sand, cobbles, and silt.

Qu — Undifferentiated gravel deposits

Qc — Colluvium
Heterogeneous deposits of rock detritus.

Travertine deposits
Travertine, of irregular thickness from 0 to 75 feet.

CRETACEOUS

Kmv — Mesaverde Formation
Interbedded sandstone and shale in upper part; lower part massive light-buff ledge-forming sandstone with lenticular coal beds.

Kc — Cody Shale

Kf — Frontier Formation
Thick lenticular gray sandstone, gray, brown, and carbonaceous shale, and bentonite.

Kmr — Mowry Shale
Gray and brown shale, in part siliceous with numerous bentonite beds and abundant fish scales.

Kt — Thermopolis Shale

CRETACEOUS / JURASSIC

KJcm — Cloverly and Morrison Formations
Cloverly Formation, of Early Cretaceous age, is light-gray sandstone, gray and variegated shale, and lenticular chert conglomerate. Morrison Formation, of Late Jurassic age, is dully variegated claystone and gray silty sandstone.

JURASSIC

Jsg — Sundance and Gypsum Spring Formations
Sundance Formation, of Late Jurassic age, is green and gray shale, greenish-gray glauconitic limy sandstone, and thin beds of fossiliferous limestone. Gypsum Spring Formation, of Middle Jurassic age, consists of red and gray shale, fossiliferous limestone, and gypsum; gypsum bed at base up to 100 feet thick.

TRIASSIC

ᵀRc — Chugwater Formation
Red siltstone, red shale, and red fine-grained sandstone; gypsiferous.

TRIASSIC / PERMIAN

ᵀRPdp — Dinwoody and Park City Formations
Dinwoody Formation, of Early Triassic age, is tan, gray, and red siltstone, gypsum, and dolomite. Park City Formation, of Permian age, is siliceous limestone and dolomite, nodular chert, and tan and gray shale.

PENN.

Pt — Tensleep Sandstone

PENNSYLVANIAN / MISSISSIPPIAN

PMa — Amsden Formation
Red shale with some dolomite limestone beds; some chert and hematite nodules; basal part commonly siltstone or sandstone.

MISSISSIPPIAN

Mm — Madison Limestone

MISSISSIPPIAN / DEVONIAN

MDtj — Three Forks and Jefferson Formations
Three Forks Formation, of Late Devonian and Early Mississippian age, is yellow, greenish-gray, and dark-gray dolomitic siltstone, black fissile shale, and silty dolomite. Jefferson Formation, of Late Devonian age, is fetid brown dolomite and light-gray and tan limestone; uppermost part is mottled yellowish-orange dolomite and yellowish-gray siltstone.

ORDOVICIAN

Ob — Bighorn Dolomite

CAMBRIAN

Єs — Snowy Range Formation
Gray-green shale and greenish flat-pebble conglomerate.

Єp — Pilgrim Limestone

Єgv — Gros Ventre Formation
Green micaceous shale, thin-bedded gray limestone, and limestone-pebble conglomerate.

Єf — Flathead Sandstone
Hard, ledge-forming quartzitic sandstone, becoming softer and brown speckled in upper part.

PRECAMBRIAN

pЄg — Granite rocks
Chiefly granite gneiss and granite

(Kf) Symbol in parentheses in Buffalo Bill Reservoir indicates location of outcrop of that formation observed when reservoir level was lowered 225 feet in 1941 for construction of Heart Mountain irrigation tunnel.

Applied Geology Exercise #1:
Devil's Tooth, Wyoming
15′ Geologic Map (1:62,500)

Name _____

Section _____

1. Each of the map units listed below represents an association of rock type and age common for western North America (and elsewhere in some cases). Using the Devil's Tooth Explanation and the chart on page 77, state the age and rock type each exemplifies. The first answer is supplied as an example.

   Chugwater Formation: _____Triassic red sandstone_____

   Crystalline rocks: _____

   Mowry Shale: _____

   Madison Formation: _____

   Flathead Formation: _____

   Mesaverde Formation: _____

2. What rock formation might have provided mineral grains for the Flathead Sandstone?
   _____

3. What two formations on this map are time-transgressive for more than one geologic time period? Give map symbol and name:

   _____, _____; _____, _____

4. Which formations contain bentonite beds that signal the emplacement of granitic batholiths to the west? _____, _____

5. Of the 12 Phanerozoic time periods, how many are represented here by one or more rock units?

   _____ Which periods are not represented? _____, _____

6. What do the relatively complete stratigraphic sequence, uniform thickness, and widespread distribution of area rock units tell us about the geologic history of this area?
   _____

7. Judging from the type and orientation of faults and folds, what sort of force pattern was imposed on this area? _____

8. Looking at the cross-section, how can you tell that folding is post-Cretaceous?
   _____
   _____

9. Is the Horse Center Anticline symmetrical or asymmetrical? Explain:

_____

_____

_____ What kind of fold is the Dry Creek

structure, located southwest of the Horse Center Anticline? _____

_____

10. Locate by general map area and section number(s) the oldest rocks on the map (Hint: consider structures present): _____

_____

11. Two lines of evidence indicate that the Half Moon Fault is not vertical. What are they? _____

_____

12. What sedimentary rock type makes up most of the petroleum-producing horizons here?

_____

13. What type of fossil fuel occurs in the Paleozoic rocks? _____

    In the Mesozoic? _____

14. Locate by general map area, section number, and rock unit a place were coal has apparently been found at the surface: _____

_____

15. What kinds of fossils would you expect to find in the Mesozoic coal layers?

_____

16. Locate the gypsum quarry on the map. In what formation are the gypsum beds located? _____

    _____ What were the environmental conditions of deposition? _____

_____

17. Why would areas on the map west of the Horse Center Anticline be a poor bet for coal deposits?

_____

18. What is the structural relationship between the granite and the Flathead Sandstone?

_____

    Check map explanations for the Grand Canyon (p. 196) and Llano Uplift (p. 208) for similar relationships. What do these three geologic settings have in common? _____

19. Construct a geologic cross-section along B-B'. Refer to instructions on p. 188.

Applied Geology Exercise #2
Geological Resources

ORIGIN AND OCCURRENCE OF PETROLEUM

Natural resources include land, water, minerals and energy. Land and water determine food production; minerals and energy supply the needs of industry and technology. Geological sciences actively contribute to all aspects of earth resources, but are particularly used to search for energy and mineral resources. Fossil energy resources largely consist of organic carbon trapped in the earth before it could be oxidized. In effect, it is "fossil sunlight," accumulated and concentrated through long periods of geologic time.[1]

Nearly half of the industrial application of geology involves the discovery and production of petroleum and natural gas. All petroleum products are relatively simple combinations of carbon and hydrogen, joined in chain or ring structures. The simplest form is methane, CH_4, which consists of a single carbon atom surrounded by four hydrogen atoms. From this simple compound, schematically represented as

$$H-\overset{H}{\underset{H}{C}}-H,$$

longer chains are formed by adding to the number of carbon atoms and corresponding hydrogens, thus

$$H-\overset{H}{\underset{H}{C}}-\overset{H}{\underset{H}{C}}-\ldots\overset{H}{\underset{H}{C}}-H.$$

The longer the chains, the heavier the petroleum products that are created. From this, a generalized formula for petroleum products may be stated as C_nH_{2n+2}.

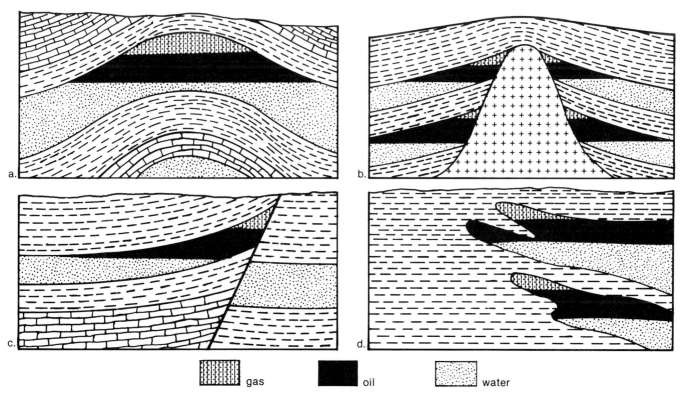

Figure 9-1. Some of the significant types of hydrocarbon traps that collect subsurface migrating oil and gas to accumulate in sufficient amount to be economically feasible for recovery. a. anticlinal trap, b. salt dome, c. fault trap, d. stratigraphic facies trap.

---

[1] Recent chemical analysis of carbon molecules which can distinguish between organic and inorganic carbon has confirmed the biologic origin of petroleum and natural gas.

Crude oil consists of a mixture of many different lengths of hydrocarbon chains, usually dominated by the heavier compounds. When crude oil is refined, it is separated according to the weight of its molecular chains by a process of simple distillation. Lighter combinations (short chains) come out as gas; heavier chains as gasoline, kerosene, fuel oil, bunker oil, or asphalt. Many years ago it was decided that the eight-chain weight was the most appropriate for motor fuel and its characteristics become known as the "octane rating" (*octo* = eight). As demand increased for automobile fuel and decreased for locomotive bunker fuel, ways were invented for breaking down the longer hydrocarbon chains into the more economically desirable shorter chains. The process is called "cracking" and is based on subjecting the long chain compounds to high temperatures, often in the presence of a catalyst, to produce shorter chains.

Petroleum is generated in organic-rich sediment and then usually flows through permeable strata until it is trapped by a structural or depositional formation. Petroleum geologists identify potential source rocks on the basis of their content of organic material (usually four to six percent for shale; less for carbonate rocks), then try to determine whether the source rocks have been subjected to enough heat to "cook" the organic material into petroleum. The temperature "window" for the formation of petroleum is between 50° and 150° Celsius. Above this temperature the petroleum chains tend to break down into natural gas. Indicators of the appropriate temperatures are found in changes in the mineralogy and chemistry of the rocks. Next, reservoir rocks, those that can contain the oil produced by source rocks, must be identified. Reservoir deposits require porous spaces where oil may be stored and connections between the spaces (permeability) so that the fluid can be removed. Sandstone and fractured limestone commonly function as reservoir rocks. Finally, structural traps must be located and a site for drilling selected. The process is somewhat like looking for a small bottle of fluid hidden in a large haystack, then trying to insert a long straw into the bottle.

Common oil traps are illustrated in Figure 9-1.

Applied Geology Exercise #2:
Origin and Occurrence of Petroleum

Name_____

Section _____

QUESTIONS

1. Why is there always an impermeable rock layer (such as shale) above and below the reservoir rock?

2. On the geologic map of Devil's Tooth, Wyoming, in the previous exercise identify the kinds of traps that have been drilled for oil. _____
   _____

3. What kinds of structures would you infer to be present at the well sites of:

   Husky Oil Company? _____

   Sunray and Mid-Continent Company? _____

4. On the reflection seismogram of a salt dome shown below mark the areas where petroleum may be trapped.

Figure 9-2. Salt Dome Reflection Seismogram

5. What kinds of problems might occur by pumping oil from underneath a heavily populated area? Discuss means of dealing with these problems.

Applied Geology Exercise #3
Ground Water

GROUND WATER RESOURCES

The amount of water on our planet is essentially a fixed quantity. Water cycles through the atmosphere, lakes, rivers, and the ocean in a predictable and reasonably constant system. Surface water is a known resource whose management is related to politics and engineering. Underground water, however, is a geological resource that must be discovered. The geological discipline which deals with ground water is **hydrogeology**. Although only six-tenths of one percent of the world's water percolates underground, its total volume is twenty times greater than that of all lakes and rivers. Many parts of the world are totally dependent on wells and springs for their water supply. In the United States, approximately 20 percent of our fresh water is supplied from natural underground reservoirs.

Water enters the ground from rainfall and temporary lakes and streams, filtering through permeable rocks or sediment (the aquifer), until it reaches an impermeable layer (the aquiclude). Permanent lakes and streams represent the level of saturated ground, called the water table. Most ground water is found in aquifers made up of rock fragments such as sandstone and conglomerate, or in carbonate rocks such as limestone. The difference between the two kinds of aquifers is significant because it affects the quantity and quality of water resources. Limestone is dissolved by water, forming solution cavities and caverns in the rock. These openings serve as pipelines and reservoirs for the movement and storage of underground water. Carbonate ground-water systems tend to fill up and drain rapidly and do not provide very effective filtering or cleansing action. Permeable sandstones or conglomerates act as immense filtering bodies through which water moves slowly, sometimes remaining in the system for thousands of years. Such water supplies are replenished slowly and are vulnerable to overproduction, but tend to hold water during droughts and to scrub many pollutants. Pumping underground water lowers the water table around a well site, forming a cone of depression. This depression lowers the water table around the well site and may cause the water to change its direction of flow.

At times an aquifer may lie between two aquicludes that prevent dispersion of the water above the water-bearing strata. In this case, water pressure caused by the downhill slope of the confined aquifer (Figure 9-3) will create a self-flowing, or artesian, well. These are usually found in valleys filled with loose sediments eroded from nearby mountains. This kind of well, like the oil gusher, is rapidly disappearing simply because it is so easy to find and exploit.

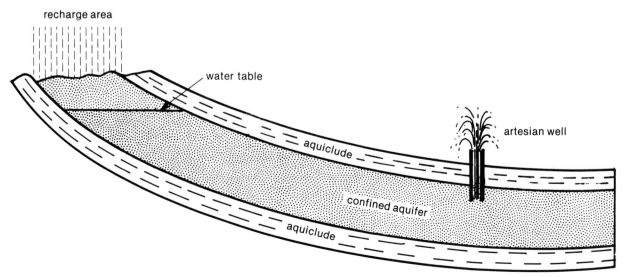

Figure 9-3. Confined Aquifer

264 Exploring Geology

Finally, a note about unpredictable ground-water distribution. No sedimentary deposit is completely uniform and predictable; surfaces that separate strata are not entirely flat. Minor hills and depressions may become filled with aquifer sediments. Even granite may have hidden internal fractures that fill with water and may be accessible. Some of these occurrences can be detected by sophisticated exploration methods, but most remain in the realm of chance discovery, located by the presence of springs or concentrated vegetation.

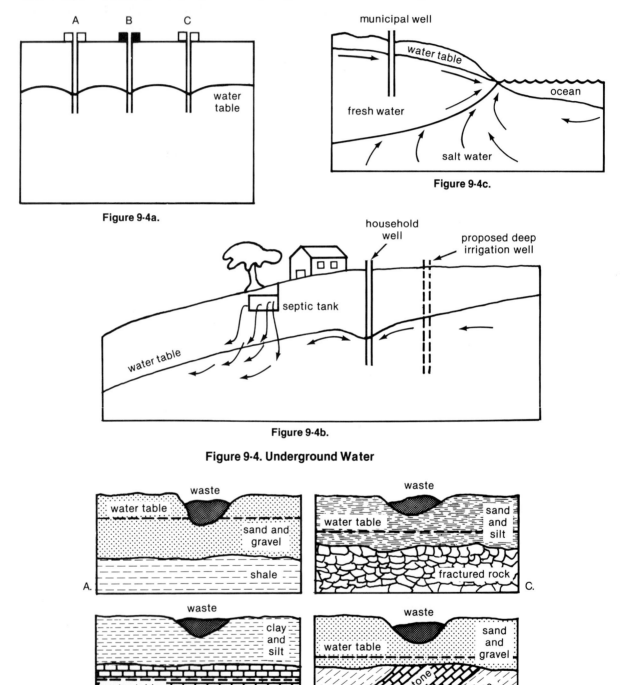

Figure 9-4. Underground Water

Figure 9-5. Effects of waste disposal on ground water. Only in B is contamination prevented by impermeable rocks. In all other examples, contaminants from the waste move through permeable rock to the water table. (From U. S. Geological Survey)

Applied Geology Exercise #3:
Ground Water

Name_____

Section _____

QUESTIONS

1. If the owner of well B in Figure 9-4a were to put in a more powerful pump to double his withdrawal of water, what would be the effect on wells A and C?

2. If the property owner in Figure 9-4b decided to put in an irrigation well deeper than the household well, what possible effect might that have on the quality of drinking water and the family's health?

3. A heavy storm washes contaminants into the aquifer that provides municipal drinking water. Compare the differences in the effect and duration of the contaminants if the aquifer were sandstone or limestone.

4. A coastal resort town depends on wells (Figure 9-4c) for its fresh water. The town grows rapidly, doubling both business and population. Suddenly all of the town's well start pumping salt water. Explain what has happened. Can the wells be recovered? If so, how?

5. A major industrial plant requests permission to bury contaminated waste (See Figure 9-5). What factors should be considered for long-term disposal? If the material is radioactive, are there other factors to be considered? Discuss.

**Applied Geology Exercise #4
Development of a Marine
Recreational Harbor**

MISSION BAY PARK,
SAN DIEGO, CALIFORNIA

PHOTOREVISED 1975 (Shown in purple)

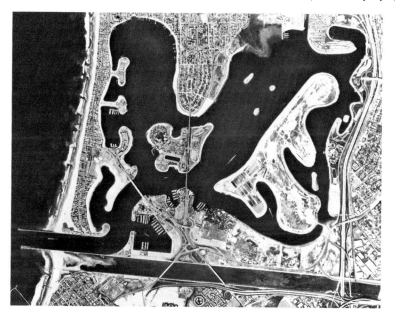

QUADRANGLE LOCATION

**Mission Bay,
San Diego, California**

This recent vertical aerial photograph shows Mission Bay Park in San Diego, California. Extensive dredge and fill operations, plus construction, have greatly modified the original sloughs here. This 4245-acre facility is now the nation's largest aquatic city park, used by over 5 million visitors a year.
(U.S.G.S. Photo)

A century ago Mission Bay was a naturally protected slough and tidal flat area, bounded on the west by a sandy baymouth bar. The undammed San Diego River brought in considerable fresh water, and built a delta (Figs. 9-6 and 9-7). Silt, contributed by the river and other drainages, filled in the bay behind the bar to an average depth at high tide of between 2 and 3 feet. Salt water moved in and out of the bay with the tides; occasional violent storms overtopped the baymouth bar. Mud-burrowing invertebrates, waterfowl, breeding fish and an occasional confused whale inhabited the area.

In the 1950s, alterations were begun which would transform these tidal flats into an extensive marine recreational facility. Dredging and filling, jetty and bridge construction, flood control, and the addition of clean sand beaches and tourist facilities drastically changed the bay's morphology and function.

Mission Bay occupies a low area which is partly a structural syncline and partly a Pleistocene erosional extension of the San Diego River. The Quaternary Rose Canyon Fault, which bounds the bay to the east, has been implicated in downwarping the bay and uplifting areas to the north and south. The recency of this fault, its relation to regional structure, and its potential for future earthquake-damage have aroused lively local discussion.

Maps in Figures 9-6 and 9-7, made 47 years apart, show the natural coast prior to human modification. The 1857 map was primarily made to aid navigation, hence its emphasis on water depth, with solid gray for shallow submerged (or shoal) areas. The full-page color map is a portion of the La Jolla, CA, 1:24,000 topographic quadrangle. It portrays the essentially completed Mission Bay Park aquatic facility. A comparison of these three maps will also serve to demonstrate how the graphic presentation of land, water, and cultural features—cartography—has evolved over 130 years.

268 Exploring Geology

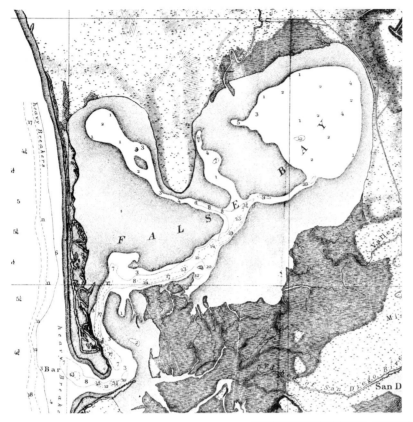

Figure 9-6
1857 Coast Survey Office Map, scale: 1:40,000. Soundings are in feet below sea level. Uniform gray areas are emergent only at low tides. A dirt road is the only culture shown.

Figure 9-7
1901-02 U.S.G.S. 15' topographic sheet, scale: 1:62,500. Culture updated 1930. Contour-like lines in water areas are decorative and unrelated to water depth.

Applied Geology 271

Applied Geology Exercise #4: Name_____
Mission Bay Park, San Diego, CA
Section _____

QUESTIONS

1. How would the present salinity of Mission Bay compare with that of the open ocean? _____

    Do you think that Mission Bay is more or less salty than it was 100 years ago? _____

    Explain: _____

    _____

2. There remains today one unmodified ½-mile stretch of Mission Bay shoreline. Where is it located?

    _____

3. Mission Beach, like most baymouth bars, has been built by the deposition of sand travelling parallel to the coast in the surf zone. Such bars are tied to the land nearest their sand supply. What is the

    direction of sand transport here? From _____ to _____

4. What is the most likely major function of the pair of jetties at the tidal inlet south of Point Medanos?

    _____

    _____

5. Why has the south end of Mission Beach grown wider since the jetties were installed? _____

    _____

6. Crown Point (Bay Point in Fig. 9-7) is the remnant of a former baymouth bar which once protected the bay from surf activity before Mission Beach existed. What two possible processes could be

    responsible for shifting the baymouth bar to the west? _____

    or _____. Comparing elevations of the two sides, state which of the

    two processes has been at work here: _____

7. The marine sedimentary rock at Crown Point has been dated at 120,000 years; the rocks were deposited between sea level and −20 feet. What is the maximum rate of vertical uplift here? Show calculations.

    _____ in./year

    The horizontal migration of the shoreline has proceeded at an average annual rate of

    _____ ft./year

8. If north-south compression has been responsible for creating the Mission Bay syncline, what is the orientation of the fold axis? _____

9. Why was Mission Bay originally called False Bay?

   _____

   _____

10. What is the source of the sediment on which Duckville was located (Fig. 9-7)? _____

    What is the name for this depositional feature? _____

11. Is the bay shallower or deeper now than it was 100 years ago? _____ If the area had been left in its natural state, would it be shallower or deeper now than 100 years ago? _____

12. Describe the basic changes in the San Diego River and its outlet which have resulted from area development (including upstream dams): _____

    _____

    _____

13. Comment briefly on the development of the travel corridor along the east side of the bay from 1857 to 1975: _____

    _____

14. What are the recreational advantages of creating lobed points, such as Santa Clara Point, in the bay?

    _____

    _____

15. List the recreational activities/facilities that exist in this area today:

    _____   _____

    _____   _____

16. List the regular urban facilities/functions of the Mission Bay area:

    _____   _____

    _____   _____

17. Reflecting on the differences between Mission Bay 100 years ago and now, do you think the transformation of this slough into a marine recreational harbor was the right thing to do? _____

    Defend your position: _____

    _____

    _____

Applied Geology Exercise #5
Urban Landsliding

THE PORTUGUESE BEND LANDSLIDE

Portuguese Bend, CA
This 1921 oblique aerial photo shows the presence of ancient landslides in the Palos Verdes area (prior to development in the 1950s). The active Portuguese Bend landslide includes all the lighter colored ground area bisected by Portuguese Canyon. The semicircular scarp to the right of this along the coast marks the head of the Beach Club landslide. The terrain at Palos Verdes is typical of Pleistocene landslide topography.

The Portuguese Bend landslide is an actively sliding slice of the larger Palos Verdes Peninsula, which is a complex mosaic of stabilized and moving blocks encompassing 300 square miles of expensive Southern California real estate (Fig. 9-9). The peninsula rose out of the sea during Pleistocene and Holocene times as a growing anticlinal structure associated with the Palos Verdes fault, which lies between the peninsula and the rest of Los Angeles to the northeast. An anticlinal crest parallels this northwest-trending fault, which has 6,000 feet of vertical displacement. Landsliding has been occurring on the peninsula for at least the last million years.

Present-day movement of the Portuguese Bend slide was inititated in 1956 by grading and filling operations to extend Crenshaw Boulevard. 160,000 cubic yards of dirt were placed to establish a grade for the road, and within days cracks began to appear in the ground near the fill. The slide soon propagated southward to the shore and west across Portuguese Canyon until all areas bounded by the present slide were activated. Ultimately 270 acres were involved in a triangular wedge of moving sedimentary and volcanic rock up to 250 feet thick and weighing 60 million tons. 160 homes and a private club were destroyed; constant maintenance of roads and utilities persists to the present day. This expensive portion of real estate is notable for above-ground sewer and water lines, enormous loops in utility wires to adjust for constantly separating poles, and driveways that lead to vacant lots. Palos Verdes Drive has been relocated twice since 1966 to keep it within the legal right-of-way, and virtually all roads seaward of it have been destroyed. The slide is "alive," moving as much as ½ inch per day, after initial velocities of up to 4 inches per day. Total horizontal translation of the slide since 1956 has been about 300 feet.

In 1961 the County of Los Angeles was assessed $9.5 million to compensate property owners in the area for their losses, on the basis that road construction initiated the disaster. This judgment may have been fair, but did little to explain how grading triggered such widespread destruction.

Geologists and engineers have had 30 years to study the Portuguese Bend slide, and during that time some clear-cut causative factors have been established. A primary condition responsible for slope failure here is the nature of the bedrock: highly silicic shales and tuffs, many of them composed of ash-derived clays which expand with water (see Fig. 9-10). These materials are

274 Exploring Geology

Figure 9-8. Variation of average monthly slide velocity with rainfall, as indicated by continuous-reading displacement gauge

Location of the Palos Verdes Hills.

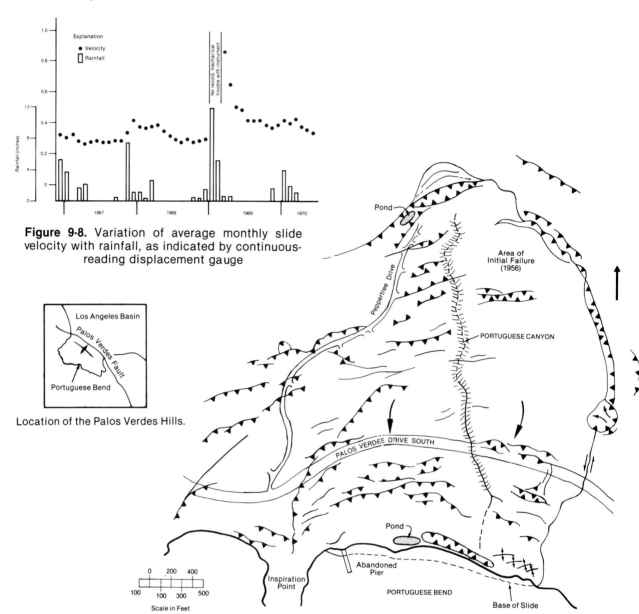

Figure 9-9. Location of the Main Portuguese Bend Landslide Area

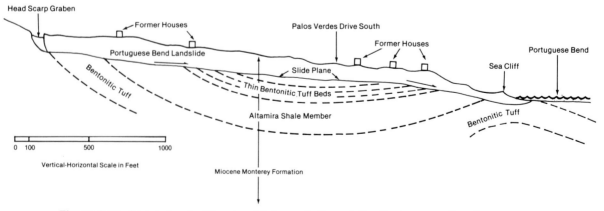

Figure 9-10. Approximate Structural Interpretation of the Portuguese Bend Landslide

weak because of well-developed bedding and low resistance to stress, especially when wet. Part of the Monterey Formation, these shales and accompanying chert, limestone and basalt units were deposited in moderately deep tropical waters of the Miocene Pacific Ocean. The rock is thoroughly shattered above the slide plane and contains many old soil horizons both above and below the active surface. These horizons are indicators of previous landslide activity, another common condition contributing to modern slides.

In addition to incompetent bedrock and pre-existing landslides, another cause for the slide is the driving force of ground water. During 1966-70 a continuous-reading displacement gauge recorded slide movements and these readings were plotted against rainfall data (Fig. 9-8). The slide is clearly sensitive to rainfall, and responds quickly to added meteoric water. There is also considerable evidence that ground-water levels have been rising during the 30-year life of the slide due to enhanced infiltration in the cracked ground of water from sag ponds (see 9-9) and from broken water mains, which have contributed millions of gallons. An 80-acre landslide adjacent to this one was stabilized in 1981 by a dewatering program which removes millions of gallons of water a year from the slide area.

Finally, most landslides are eventually self-stabilized by the development of a toe, a natural buttress of accumulated slide debris at the low end. At Portuguese Bend, however, a permanent plume of silt beyond the surf zone attests to continual toe removal by wave action, erasing this potential barrier to further sliding before it can form.

Applied Geology 277

Applied Geology Exercise #5:
Portuguese Bend Landslide

Name _____

Section _____

QUESTIONS

1. Outlined below are several major factors which engineering geologists analyze when assessing slope stability. Comment on each as it pertains to the Portuguese Bend area.

    a. nature of bedrock _____
    _____

    b. regional structure _____
    _____

    c. presence of groundwater _____
    _____
    _____

    d. presence of prior landsliding _____
    _____

    e. local construction practice _____
    _____

    f. development of slide "toe" _____
    _____

2. What Southern California geologic hazard besides those listed in Question #1 above might also induce slides in this area? _____

3. Given the rock types present, what was going on tectonically in Miocene Southern California that is not occurring today? _____
_____

4. No one has been able to determine whether the basalt units here are flows or sills. What would *you* look for in the field to solve this question? _____
_____
_____

278 Exploring Geology

5. Much of Southern California had extensive landsliding during the Pleistocene. Excluding tectonic factors, what else was happening then that would have encouraged sliding? _____

_____

6. Portuguese Canyon has no alluvial fan where it emerges at the coast. What would this indicate about the fate of water entering the head of the canyon? _____ _____ What effect would this have on the slide? _____

_____

7. Why are sag ponds on a landslide likely to help keep the slide moving? _____

_____

8. What is the evidence for ancient landsliding in this area? _____

_____

9. Given the orientation of the Palos Verdes anticline (Fig. 9-8), what is the dip direction of the beds at the Portuguese Bend site? _____

   How would this contribute to slope instability? _____

_____

10. Los Angeles County at one point tried to halt the slide by "pinning" it with pre-cast concrete cylinders. The slides continues unabated, and many cylinders have disappeared. What do you think millions of tons of soft, dense, moving rock *did* to those "pins"? _____

_____

11. Think of one or more technically feasible ways to stabilize, or slow down, the Portuguese Bend landslide. _____

_____

_____

12. Do you think the $9.5 million judgment against the County of Los Angeles for initiating the slide was fair? _____ Defend your position: _____

_____

_____

_____

Applied Geology Exercise #6
Tectonic Hazards

LOCATING EARTHQUAKE EPICENTERS

San Andreas Fault, CA
This aerial view shows the San Andreas Fault where it crosses the Carrizo Plain in California's Central Valley. The Elkhorn escarpment is clearly visible in this photo.
(Photo by R. E. Wallace, U.S.G.S.)

Earthquakes are one of the most terrifying geologic events. They occur suddenly, usually without warning, and in less than five minutes may cause destruction and suffering on a scale difficult to comprehend. Ironically, earthquakes usually do not damage the land or even seriously change its topography, but destroy only those things which are built on top of the land. At present we cannot control earthquakes or even predict them within an acceptable time frame. Our methods of adapting to earthquakes include avoiding areas where they may occur and designing structures that will not be destroyed by them. The accompanying map gives an idea of the United States where earthquakes are most likely to occur. However, the map does not indicate probable magnitude of earthquakes. We should remember that the most powerful earthquake known in North America took place in Missouri.

The exact place within the crust of the earth where an earthquake occurs is called its **focus**. The spot on the surface of the earth directly above the focus—the surface area the shortest distance away and therefore the most affected—is called the earthquake's **epicenter**. An epicenter is located by determining the time interval between the arrival at a seismograph station of **P earthquake waves** (rapid compression waves) and **S earthquake waves** (slower shear waves). This interval can be used to calculate the distance to the earthquake, but not the direction. To establish the precise location of an epicenter, the distance of shock waves from two or more seismic stations must be plotted as arcs from each station. The epicenter is located where the arcs intersect.

280 Exploring Geology

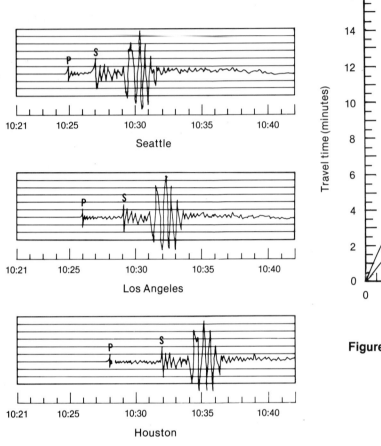

Figure 9-12. Time-Distance Graph of Seismic Waves

(Greenwich Mean Time)

Figure 9-11. Seismographic Traces

MEASURING EARTHQUAKE MAGNITUDE AND INTENSITY

Two scales are commonly used to measure earthquakes. The Richter Scale provides an accurate measure of the amount of energy released by the sudden movement of vast quantities of rock. The strength of the rocks in the earth's crust is a known quantity. By measuring the amplitude of the earthquake wave on a seismograph, and knowing the distance it has traveled, a calculation can be made that measures the energy required to move the rocks. Calculations are then expressed on a logarithmic scale, in which each number represents an increase equal to about ten times the magnitude of the previous number. Thus an earthquake of magnitude 7 is ten times more powerful than one of magnitude 6 and one hundred times more than an earthquake rated 5 on the Richter Scale. The formula for calculating Richter magnitude is as follows:

$$\text{Richter Magnitude} = \text{Log}\left(\frac{a}{T}\right) + B,$$

where **a** is the ground motion measured by the seismograph, **T** is the duration of one seismic wave, and **B** is the rock attenuation factor.

The largest earthquakes recorded have measured about 8.5 on the Richter scale; damage to structures does not begin until about magnitude 5.

Unlike the Richter Scale, which measures a quantity of energy, the Mercalli Scale provides a relative measure of the intensity of ground-shaking based on the amount and kind of damage actually done by the earthquake. This scale is more useful than the Richter Scale to city planners and engineers who must deal with the effects of earthquakes on human populations and structures. A shortened version of the Mercalli Scale (modified from Richter 1958) is as follows:

I. Not felt. Marginal and long-period effects of large earthquakes.

II. Felt by persons at rest, on upper floors, or favorably placed.

III. Felt indoors. Hanging objects swing. Vibration like passing of light trucks. Duration estimated. May not be recognized as an earthquake.

IV. Hanging objects swing. Vibration like passing of heavy trucks; or sensation of a jolt like a heavy ball striking the walls.

V. Felt outdoors; direction estimated. Sleepers wakened. Liquids disturbed, some spilled. Small unstable objects displaced or upset. Doors swing, close, open. Shutters, pictures move.

VI. Felt by all. Many frightened and run outdoors.* People walk unsteadily. Windows, dishes, glassware broken. Furniture moved or overturned. Weak plaster and masonry D (weak masonry) cracked. Small bells ring (church, school).

VII. Difficult to stand. Noticed by drivers of motor cars. Hanging objects quiver. Furniture broken. Damage to masonry D, including cracks. Weak chimneys broken at roof line. Fall of plaster, loose bricks, stones, tiles, cornices, unbraced parapets, and architectural ornaments.

VIII. Steering of automobiles affected. Damage to masonry C (ordinary masonry); partial collapse of masonry D. Some damage to masonry B (good masonry); none to masonry A (excellent masonry). Fall of stucco and some masonry walls. Twisting or falling of chimneys, factory smokestacks, monuments, towers, elevated tanks.* Changes in flow or temperature of springs and wells.

IX. General panic.* Masonry D destroyed; masonry C heavily damaged, sometimes with complete collapse; masonry B seriously damaged. General damage of foundations. Frames cracked. Serious damage to reservoirs.* Underground pipes broken.* Conspicuous cracks in ground.* In alluviated areas sand and mud ejected, earthquake fountains, sand craters.*

X. Most masonry and frame structures destroyed with their foundations. Some well-built wooden structures and bridges destroyed.* Serious damage to dams, dikes, embankments.* Large landslides.*

XI. Rails bent greatly.* Underground pipelines completely out of service.*

XII. Damage nearly total. Large rock masses displaced.* Lines of sight and level distorted.* Objects thrown into the air.*

*These criteria may be misleading as a measure of the strength of shaking.

Applied Geology Exercise #6:
Tectonic Hazards

Name _____

Section _____

QUESTIONS

1. Using the Time-Distance Graph and the Seismographic Traces on page 280, complete the information below and then plot the epicenter of the earthquake on the map. Remember that it is the time interval between the P- and S-wave travel time that determines the distance.

    Data from Seismograph

    Seattle: P-S interval _____ minute(s).

    Los Angeles: P-S interval _____ minute(s).

        First P-wave arrives _____ minute(s) after Seattle.

    Houston: P-S interval _____ minute(s).

        First P-wave arrives _____ minute(s) after Seattle.

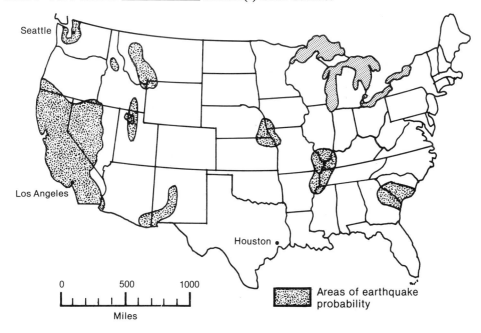

2. Now that you know how far the epicenter is from each seismograph station and the arrival time of the first P-wave at each station calculate the exact time of the earthquake by using the Time-Distance Graph to determine when the P-wave left the epicenter.

    Time of earthquake _____

3. On a world-wide basis where do most earthquakes occur today, and why?

4. Shown on the map above are areas of *likely earthquake probability*. Based on your knowledge of plate tectonics give some reasons why these areas are the most likely places for earthquakes.

284 Exploring Geology

5. What can architects and urban planners do to lessen the damage from earthquakes? How can geologic information assist in these decisions?

6. How would you determine the difference between a minor earthquake and a major nuclear bomb test? Discuss.

Applied Geology Exercise #7
Mount St. Helens, Washington

INTRODUCTION TO MOUNT ST. HELENS VOLCANO, WASHINGTON

Mount St. Helens, WA
Photo shows the eruption of Mount St. Helens on May 18, 1980. Note the immense ash and dust cloud as well as the pyroclastic rock and mud river flowing north towards the right side of the photo. (U.S.G.S. Photo)

On May 18, 1980 after 58 days of increasing earthquake activity, steam and ash release, Mount St. Helens erupted in a cataclysmic volcanic event which claimed more than 60 human lives and altered the landscape for millenia to come. The financial loss was in excess of one billion dollars. Hundreds of houses and billions of board feet of salable timber were destroyed. Thousands of animals and millions of fingerling salmon were also killed in the eruption and its aftermath.

Could this eruption have been prevented? No. Did we know it was coming? Yes. Hazard warnings and restricted access saved more lives than were lost. Scientists have learned a great deal about the nature of composite volcanos as hazards and as geological entities. This eruption of Mount St. Helens is the best monitored and documented major volcanic event in human history.

Prior to 1980 Mount St. Helens had the most nearly perfect conical form of all the 15 Cascade composite (stratoform) volcanos—a sure sign that volcanic activity had been geologically recent and that the landform was young. Frequent Pleistocene eruptions held to a minimum the erosive effects of glaciers which scarred neighboring volcanos Mt. Adams and Mt. Rainier. In 1978 Dwight Crandell and Donal Mullineaux of the U.S.G.S. authored a Survey Bulletin entitled *Potential Hazards from Future Eruptions of Mount St. Helens Volcano, Washington* which anticipated a near-future eruption and its consequences with startling accuracy.

Earthquake activity in March 1980 signalled the end of a 120-year dormant period for the volcano (Fig. 9-13a). The north side of the mountain began to bulge ominously (Fig. 9-13c) as magma accumulated within. The catastrophic sequence of events on May 18 began with a Richter magnitude 5 earthquake. This triggered the largest debris landslide recorded in historic time as multiple slices of the unstable north side gave way to gravity and avalanched 13 miles down the North Fork Fork of the Toutle River valley filling it to a depth of 150 feet. It also raised the bottom of Spirit Lake by 295 feet. This large-scale lateral sliding "uncorked" the pressurized magma chamber, initiating an explosive blast which moved at speeds up to 670 miles per hour, devastating a 230 square-mile wedge northeast of the volcano. It removed or flattened everything in its path.

This blast, which killed U.S.G.S. volcanologist David Johnston at the Coldwater II observation station, was responsible for almost all of the timber loss in the eruption. Pyroclastic flows emerged near the summit shortly after the blast as the volcano frothed over, depositing 17 pumice-rich flows. The volcano pumped 540 tons of ash into the atmosphere in the ensuing nine hours. While most of the ash settled on 22,000 square miles of several western states, some of the finer material that encircled the globe within days of the eruption is still aloft today. Loose debris ejected from the volcano mixed with melted snowpack and river water to create enormous fast-moving mudflows. 65 million cubic yards of saturated ejecta hurtled down the Cowlitz and Columbia Rivers at speeds of up to 90 miles per hour, wiping out bridges and roads and filling up reservoirs and shipping lanes. River water levels swelled to flood stage of 21 feet above normal. Figure 9-13b shows the areal distribution of the various large-scale eruptive deposits and effects.

Since the catastrophic activity in 1980, about 20 smaller eruptions have taken place. These involve steam and ash emission, and the cyclic growth and partial destruction of viscous lava plugs (domes) within the crater. Should this continue, this slow growth of the dome will require nearly a century to rebuild the pre-1980 conical summit.

Applied Geology 287

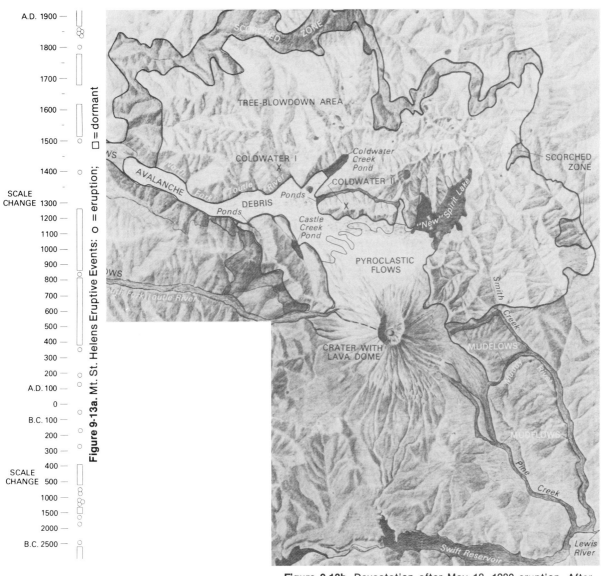

Figure 9-13a. Mt. St. Helens Eruptive Events: o = eruption; ▢ = dormant

Figure 9-13b. Devastation after May 18, 1980 eruption. After Molenaar, U.S.G.S. Professional Paper 1249, 1982.

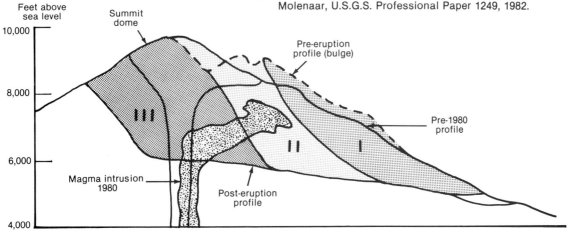

Figure 9-13c. Generalized N-S cross-section showing magma intrusion and the three blocks which collapsed to form the debris avalanche (block I slid first, etc.) May 18, 1980.

Figure 9-13. Mt. St. Helens

Applied Geology Exercise #7:
Mount St. Helens Geology and Hazards

Name _____

Section _____

QUESTIONS

1. Estimate the maximum vertical swelling (in feet) of the 1980 pre-eruption bulge. _____

2. What triggered the powerful lateral blast and lava extrusion?
   _____

3. What topographic areas were most vulnerable to mudflows? _____
   _____

4. Volcanic ash is composed of microscropic, razor-sharp shards of volcanic glass. What problems would this ash cause residents in the outfall area?
   _____
   _____

5. Briefly describe the tectonic setting of the Cascade composite volcanos. (See pages 228-229.)
   _____
   _____

6. Make a generalization about Mount St. Helens' dormant periods since A.D. 0.
   _____
   _____

7. Considering the effect of the eruption on Spirit Lake, what additional hazard may be associated with the lake in the next few years? _____
   _____

8. Prior to 1980 what evidence existed that Mount St. Helens was among the most active volcanos of the Cascade Range? _____
   _____

9. What geologic conditions or events preceded the May 18 event and warned of the forthcoming eruption? _____
   _____
   _____

10. What aspect of the eruption was responsible for each of these events?

    a. downstream flooding _____

    b. most of the timber loss _____

    c. raising Spirit Lake bottom _____

    d. respiratory problems_____

11. The lava flows emitted during the May 18 eruption were mostly of the volcanic rock type _____.

12. The thick, sticky nature of the post-eruptive volcanic dome indicates what general igneous rock composition? _____

    _____

13. Considering the geographic location of Mt. St. Helens (see page 228) why might a southward-directed blowout have been far more costly and dangerous than the northward-directed one which occurred?

    _____

14. Name two other Cascade volcanos which are situated uncomfortably close to population centers (see Fig. 8-4, page 228).

    _____ _____

15. Consult your lecture text or other references to discover the only other volcano in the contiguous 48 United States to have erupted this century. _____

16. Long Valley Caldera (Mammoth Lakes, CA) and Yellowstone Caldera (Wyoming) ejected about 100 times as much material as Mt. St. Helens. Which Cascade volcano probably had a similar volume of ejecta? Explain your answer. _____

    _____

    _____

    _____

Index

A

Aerial photographs, 57-69
       parallax in, 59
       photomosaic using, 58
       types of, 59-60
Anticline, 147, 149, 183

B

Bacon, Sir Francis, 215
Basalt, 37-38, 48-49
Batholith, 35-36
Base line and meridian systems, 99-101

C

Caldera, 225
Canadian Shield, 30
Cascade Range, 30, 228-29
Columbia Plateau, 30
Composite volcanos, 37
Concrete, 29
Contact, 182-83, 188
Cross-sections, 181, 184-86, 188-89

D

Death Valley, 29
Dike, 35

E

Earthquakes, 279-283

F

Faults, 150-53, 188
Ferromagnesian minerals, 6
Fissure flows, 37
Folds, 143-44, 147-49
Fool's gold (pyrite), 2, 7, 24
Formations, 181
Fossils, 40-41, 71-73, 75, 80-83, 85, 87

G

Graben, 150-151
Gradient, 104
Grand Canyon, 39, 181, 195-202
Great Lakes, 39
Ground water, 263
Gulf Coast, 39

H

Hardness scale, Mohs', 2
Hawaii, 29, 234
Holmes, Oliver Wendell, 57

Horsts, 150, 151
"Hot spots," 217

I

Isopachs, 241

J

Joints, 149

L

Laccolith, 35
Landsliding, 273-75
Latitude and longitude, 98-99
Lava, 30, 37, 116, 215, 225, 229
Linné, Carl von, 71
Linnaean classification, 71-73
Lithosphere, dynamic, 48, 215

M

Magma, 29-31
Maps
       contours in, 102-103, 107
       geologic, 181-207
       grids of, 98
       relief in, 58, 97, 104, 107
       projections used in, 97-98
       planimetric, 97
       scales in, 102
       topographic 97-142
Minerals, 1-28
       kinds of, 6-13
       luster of, 3, 17-19
       properties of, 3-5
Monocline, 147, 183
Mount St. Helens, cover, 228-29, 285-90

O

Oceans, physiography, 219-23
Ore, 6

P

Pacific Ring of Fire, the, 37, 228
Petroleum, composition and production, 259-60
Photogrammetry, 57
Plutonic rocks, 35-36
Pyroclastic, 37

R

Resources, geological, 259
Rift valleys, 239

Rocks, 29-56
    chemical and biochemical compared, 39, 43-44
    defined, 29
    deformations of, 143, 147
    detrital, 39, 41-42
    dynamic lithosphere and, 48
    features of, 39-40
    igneous, 30-38
    metamorphic, 45-47
    sedimentary, 39-44
    types of, 49-50
Rule of V's, the, 185, 187, 190

S

San Andreas Fault, the, 234, 247
San Francisco, CA, 64, 66, 234-35
Scarps, 151
Sediment, 39
Sill, 35
Slickensides, 150
Stock, 35
Strike and dip, 146, 155, 157-160, 164, 165, 177, 183
Sutures, 217
Syncline, 147, 148-149, 183

T

Tectonics, 48
    plate, 215-250
Terranes, 215
Township and range, 99-101

U

Unconformity, 154, 156

Long./Lat.
Map symbols chart
bring water ever you want
Conversion Miles-feet/km
Range-township no.
· ~~PLS~~ number

12:30 Wednes. Nov 6
1hr - Topo. Review